Lecture Notes in Mathematics

Edited by A. Dold and B. Eckmann

427

Hideki Omori

Infinite Dimensional Lie Transformation Groups

Springer-Verlag
Berlin · Heidelberg · New York

Lecture Notes in Mathematics

Vols. 1-183 are also available. For further information, please contact your book-seller or Springer-Verlag.

Vol. 184: Symposium on Several Complex Variables, Park City, Utah, 1970. Edited by R. M. Brooks. V, 234 pages. 1971. DM 20,–

Vol. 185: Several Complex Variables II, Maryland 1970. Edited by J. Horváth. III, 287 pages. 1971. DM 24,–

Vol. 186: Recent Trends in Graph Theory. Edited by M. Capobianco/ J. B. Frechen/M. Krolik. VI, 219 pages. 1971. DM 18,–

Vol. 187: H. S. Shapiro, Topics in Approximation Theory. VIII, 275 pages. 1971. DM 22,–

Vol. 188: Symposium on Semantics of Algorithmic Languages. Edited by E. Engeler. VI, 372 pages. 1971. DM 26,–

Vol. 189: A. Weil, Dirichlet Series and Automorphic Forms. V, 164 pages. 1971. DM 16,–

Vol. 190: Martingales. A Report on a Meeting at Oberwolfach, May 17-23, 1970. Edited by H. Dinges. V, 75 pages. 1971. DM 16,–

Vol. 191: Séminaire de Probabilités V. Edited by P. A. Meyer. IV, 372 pages. 1971. DM 26,–

Vol. 192: Proceedings of Liverpool Singularities – Symposium I. Edited by C. T. C. Wall. V, 319 pages. 1971. DM 24,–

Vol. 193: Symposium on the Theory of Numerical Analysis. Edited by J. Ll. Morris. VI, 152 pages. 1971. DM 16,–

Vol. 194: M. Berger, P. Gauduchon et E. Mazet. Le Spectre d'une Variété Riemannienne. VII, 251 pages. 1971. DM 22,–

Vol. 195: Reports of the Midwest Category Seminar V. Edited by J. W. Gray and S. Mac Lane. III, 255 pages. 1971. DM 22,–

Vol. 196: H-spaces – Neuchâtel (Suisse)- Août 1970. Edited by F. Sigrist. V, 156 pages. 1971. DM 16,–

Vol. 197: Manifolds – Amsterdam 1970. Edited by N. H. Kuiper. V, 231 pages. 1971. DM 20,–

Vol. 198: M. Hervé, Analytic and Plurisubharmonic Functions in Finite and Infinite Dimensional Spaces. VI, 90 pages. 1971. DM 16,–

Vol. 199: Ch. J. Mozzochi, On the Pointwise Convergence of Fourier Series. VII, 87 pages. 1971. DM 16,–

Vol. 200: U. Neri, Singular Integrals. VII, 272 pages. 1971. DM 22,–

Vol. 201: J. H. van Lint, Coding Theory. VII, 136 pages. 1971. DM 16,–

Vol. 202: J. Benedetto, Harmonic Analysis in Totally Disconnected Sets. VIII, 261 pages. 1971. DM 22,–

Vol. 203: D. Knutson, Algebraic Spaces. VI, 261 pages. 1971. DM 22,–

Vol. 204: A. Zygmund, Intégrales Singulières. IV, 53 pages. 1971. DM 16,–

Vol. 205: Séminaire Pierre Lelong (Analyse) Année 1970. VI, 243 pages. 1971. DM 20,–

Vol. 206: Symposium on Differential Equations and Dynamical Systems. Edited by D. Chillingworth. XI, 173 pages. 1971. DM 16,–

Vol. 207: L. Bernstein, The Jacobi-Perron Algorithm – Its Theory and Application. IV, 161 pages. 1971. DM 16,–

Vol. 208: A. Grothendieck and J. P. Murre, The Tame Fundamental Group of a Formal Neighbourhood of a Divisor with Normal Crossings on a Scheme. VIII, 133 pages. 1971. DM 16,–

Vol. 209: Proceedings of Liverpool Singularities Symposium II. Edited by C. T. C. Wall. V, 280 pages. 1971. DM 22,–

Vol: 210: M. Eichler, Projective Varieties and Modular Forms. III, 118 pages. 1971. DM 16,–

Vol. 211: Théorie des Matroïdes. Edité par C. P. Bruter. III, 108 pages. 1971. DM 16,–

Vol. 212: B. Scarpellini, Proof Theory and Intuitionistic Systems. VII, 291 pages. 1971. DM 24,–

Vol. 213: H. Hogbe-Nlend, Théorie des Bornologies et Applications. V, 168 pages. 1971. DM 18,–

Vol. 214: M. Smorodinsky, Ergodic Theory, Entropy. V, 64 pages. 1971. DM 16,–

Vol. 215: P. Antonelli, D. Burghelea and P. J. Kahn, The Concordance-Homotopy Groups of Geometric Automorphism Groups. X, 140 pages. 1971. DM 16,–

Vol. 216: H. Maaß, Siegel's Modular Forms and Dirichlet Series. VII, 328 pages. 1971. DM 20,–

Vol. 217: T. J. Jech, Lectures in Set Theory with Particular Emphasis on the Method of Forcing. V, 137 pages. 1971. DM 16,–

Vol. 218: C. P. Schnorr, Zufälligkeit und Wahrscheinlichkeit. IV, 212 Seiten. 1971. DM 20,–

Vol. 219: N. L. Alling and N. Greenleaf, Foundations of the Theory of Klein Surfaces. IX, 117 pages. 1971. DM 16,–

Vol. 220: W. A. Coppel, Disconjugacy. V, 148 pages. 1971. DM 16,–

Vol. 221: P. Gabriel und F. Ulmer, Lokal präsentierbare Kategorien. V, 200 Seiten. 1971. DM 18,–

Vol. 222: C. Meghea, Compactification des Espaces Harmoniques. III, 108 pages. 1971. DM 16,–

Vol. 223: U. Felgner, Models of ZF-Set Theory. VI, 173 pages. 1971. DM 16,–

Vol. 224: Revètements Etales et Groupe Fondamental. (SGA 1). Dirigé par A. Grothendieck XXII, 447 pages. 1971. DM 30,–

Vol. 225: Théorie des Intersections et Théorème de Riemann-Roch. (SGA 6). Dirigé par P. Berthelot, A. Grothendieck et L. Illusie. XII, 700 pages. 1971. DM 40,–

Vol. 226: Seminar on Potential Theory, II. Edited by H. Bauer. IV, 170 pages. 1971. DM 18,–

Vol. 227: H. L. Montgomery, Topics in Multiplicative Number Theory. IX, 178 pages. 1971. DM 18,–

Vol. 228: Conference on Applications of Numerical Analysis. Edited by J. Ll. Morris. X, 358 pages. 1971. DM 26,–

Vol. 229: J. Väisälä, Lectures on n-Dimensional Quasiconformal Mappings. XIV, 144 pages. 1971. DM 16,–

Vol. 230: L. Waelbroeck, Topological Vector Spaces and Algebras. VII, 158 pages. 1971. DM 16,–

Vol. 231: H. Reiter, L^1-Algebras and Segal Algebras. XI, 113 pages. 1971. DM 16,–

Vol. 232: T. H. Ganelius, Tauberian Remainder Theorems. VI, 75 pages. 1971. DM 16,–

Vol. 233: C. P. Tsokos and W. J. Padgett. Random Integral Equations with Applications to stochastic Systems. VII, 174 pages. 1971. DM 18,–

Vol. 234: A. Andreotti and W. Stoll. Analytic and Algebraic Dependence of Meromorphic Functions. III, 390 pages. 1971. DM 26,–

Vol. 235: Global Differentiable Dynamics. Edited by O. Hájek, A. J. Lohwater, and R. McCann. X, 140 pages. 1971. DM 16,–

Vol. 236: M. Barr, P. A. Grillet, and D. H. van Osdol. Exact Categories and Categories of Sheaves. VII, 239 pages. 1971. DM 20,–

Vol. 237: B. Stenström, Rings and Modules of Quotients. VII, 136 pages. 1971. DM 16,–

Vol. 238: Der kanonische Modul eines Cohen-Macaulay-Rings. Herausgegeben von Jürgen Herzog und Ernst Kunz. VI, 103 Seiten. 1971. DM 16,–

Vol. 239: L. Illusie, Complexe Cotangent et Déformations I. XV, 355 pages. 1971. DM 26,–

Vol. 240: A. Kerber, Representations of Permutation Groups I. VII, 192 pages. 1971. DM 18,–

Vol. 241: S. Kaneyuki, Homogeneous Bounded Domains and Siegel Domains. V, 89 pages. 1971. DM 16,–

Vol. 242: R. R. Coifman et G. Weiss, Analyse Harmonique Non-Commutative sur Certains Espaces. V, 160 pages. 1971. DM 16,–

Vol. 243: Japan-United States Seminar on Ordinary Differential and Functional Equations. Edited by M. Urabe. VIII, 332 pages. 1971. DM 26,–

Vol. 244: Séminaire Bourbaki – vol. 1970/71. Exposés 382–399. IV, 356 pages. 1971. DM 26,–

Vol. 245: D. E. Cohen, Groups of Cohomological Dimension One. V, 99 pages. 1972. DM 16,–

Vol. 246: Lectures on Rings and Modules. Tulane University Ring and Operator Theory Year, 1970–1971. Volume I. X, 661 pages. 1972. DM 40,–

continuation on page 155

Lecture Notes in Mathematics

Edited by A. Dold and B. Eckmann

427

Hideki Omori

Infinite Dimensional Lie Transformations Groups

Springer-Verlag
Berlin · Heidelberg · New York 1974

Prof. Hideki Omori
Tokyo Metropolitan University
Fukazawa Setagaya
Tokyo/Japan

Library of Congress Cataloging in Publication Data

Omori, Hideki, 1938-
Infinite dimensional Lie transformation groups.

(Lecture notes in mathematics ; 427)
Bibliography: p.
Includes index.
1. Transformation groups. 2. Lie groups.
3. Manifolds (Mathematics) I. Title. II. Series:
Lecture notes in mathematics (Berlin) ; 427.
QA3.L28 no. 427 [QA274.7] 512'.55 74-23625

AMS Subject Classifications (1970): 54H15, 58B99

ISBN 3-540-07013-3 Springer-Verlag Berlin · Heidelberg · New York
ISBN 0-387-07013-3 Springer-Verlag New York · Heidelberg · Berlin

Offsetdruck: Julius Beltz, Hemsbach/Bergstr.

Introduction

In this article, the author wants to discuss the possibilities of an infinite dimensional analogue of the theory of finite dimensional Lie groups. As such an analogue, we have already theories of Banach Lie groups or Hilbert Lie groups, which are infinite dimensional analogues of finite dimensional linear groups, and under some suitable conditions of simpleness, we have already a classification table of such groups. (Cf. [7] for instance.)

Why does one want another theory of infinite dimensional Lie groups ? Because of the following facts :

a) Let G be a Banach Lie group with Lie algebra $\mathfrak{g}$. If $\mathfrak{g}$ has no proper, finite codimensional ideal, then the only possible smooth action of G on a finite dimensional smooth manifold is trivial. (Cf. [33].)

The above fact shows that Banach Lie groups rarely act on finite dimensional manifolds. We do not have a single example of infinite dimensional Banach Lie group which acts effectively and transitively on a compact manifold. In many cases, Banach Lie groups which act on finite dimensional manifolds turn out to be finite dimensional Lie groups.

b) If a Banach Lie group acts smoothly, effectively, transitively and primitively (i.e. leaves no foliation invariant) on a finite dimensional manifold, then it must be a finite dimensional Lie group. (Cf. [33].)

In contrast, Leslie [20] showed that the group of all C^∞-diffeomorphisms on a closed manifold is a Frechet Lie group, namely this group is an infinite dimensional manifold modeled on a Frechet space and the group operations are smooth.

However, general Frechet manifolds are very difficult to treat. For instance, there are some difficulties in the definition of tangent bundles, hence in the definition of the concept of C^∞-mappings. Of course, there is neither an implicit function theorem nor a Frobenius theorem in general. Thus, it is difficult to give a theory of general Frechet Lie groups. It is better to consider an intermediate concept between Banach Lie groups and Frechet Lie groups.

For this purpose, the author defines the concept of strong ILB- and strong ILH-Lie groups. Even in this category, there is neither an implicit function theorem nor a Frobenius theorem in general. However, we give a sufficient condition by using these concepts.

In §I, the precise definition of strong ILB(or ILH)-Lie groups will be given, and some general facts will be discussed. In this category of groups, one can defin Lie algebras and exponential mappings. Moreover, the following theorem holds :

Theorem A The group structure of a strong ILB-Lie group is locally determined by its Lie algebra.

Now, it is natural to ask how many strong ILB (or ILH)-Lie groups exist. So, the main purpose of this article is to find examples. We remark : The concept of strong ILB (or ILH)-Lie groups is something like the concept of structures defined on topological groups. A topological group can have many strong ILB-Lie group structures.

Throughout this article, M denotes a closed C^∞- manifold and $\mathcal{D}(M)$ denotes the group of all C^∞-diffeomorphisms of M with C^∞-topology.

Theorem B The topological group $\mathcal{D}(M)$ has both strong ILH- and strong ILB-Lie group structures.

By a slight modification of the proof of the above theorem, we get the followi

Theorem C Suppose M has a C^∞-fibering with a compact fiber. Then, the subgroup $\mathcal{D}_{\mathcal{F}}(M)$ of $\mathcal{D}(M)$ which leaves the fibering $\mathcal{F}$ invariant has both strong ILH- and ILB-Lie group structures.

Theorem D Let K be a compact subgroup in $\mathcal{D}(M)$. Then, the subgroup $\mathcal{D}_K(M)$ of $\mathcal{D}(M)$ of elements which commute pointwise with all $k \in K$ has both strong ILH- an

ILB-Lie group structures.

Theorem E Let S be a closed C^∞-submanifold of M. Let

$$\mathcal{D}(M,S) = \{ \psi \in (M) : \psi(S) = S \}$$

$$\mathcal{D}(M,[S]) = \{ \psi \in (M) : \psi(x) = x \text{ for any } x \in S \} .$$

Then, $\mathcal{D}(M,S)$ and $\mathcal{D}(M,[S])$ have strong ILB-Lie group structures as subgroups of the strong ILB-Lie group $\mathcal{D}(M)$.

The concept of strong ILB- (or ILH-) Lie subgroups will be given in §I. The above theorem can be extended in several ways. These are discussed in §II together with the proof of Theorems B - E.

In §III, we establish the implicit function theorem on strong ILB-Lie groups, and prove that strong ILB- (or ILH-) Lie group structures on $\mathcal{D}(M)$, $\mathcal{D}_{\mathcal{J}}(M)$, $\mathcal{D}_K(M)$ etc. do not depend on the choice of connections, inner products and volume elements etc.. Moreover, we have the following :

Theorem F Let G be a strong ILB-Lie group acting smoothly on a finite dimensional manifold M. Then the isotropy subgroup G_m at $m \in M$ is a strong ILB-Lie subgroup of G and the orbit $G(m)$ is a smooth submanifold of M.

Now, a strong ILB- (or ILH-) Lie group is not a special group. Moreover, it appears very often when the group is a non-linear transformation group on some space. It will be illustrated implicitly in the following :

Theorem G Suppose we have a linear representation ρ of a strong ILB- (resp. ILH-) Lie group G to a Sobolev chain $\{\mathbb{F}, F^s, s \geq d\}$ such that $\rho(gh) = \rho(h)\rho(g)$ and that the mapping $\rho : \mathbb{F} \times G \mapsto \mathbb{F}$ can be extended to the C^ℓ-mapping of $F^{s+\ell} \times G^s$ into F^s for every $s \geq d$. Then, the semi-direct product $\mathbb{F}*G$, that is, the direct product $\mathbb{F} \times G$ with the group operation $(f,g)*(f',g') = (\rho(g')f + f', gg')$, is a strong ILB- (resp. ILH-) Lie group.

We remark that even if G is a finite dimensional Lie group, F^s*G is not a Banach Lie group in general, because the group operation may not be differentiable.

The above theorem will be proved in §IV.

In §IV, we also discuss an abstract treatment of vector bundles over strong ILB-groups. Some of the results obtained here will be used in §VII.

In §V, we will give a short review of the smooth extension theorem [31] and a remark on elliptic operators. These results are combined to obtain a Frobenius Theorem in §VII.

In § VI, we will give a basic idea of the proof of Frobenius theorem. Frobenius theorem on strong ILH-Lie groups will be given in §VII.

The purpose of §VIII is to give miscellaneous examples of strong ILB- and ILH-Lie groups.

In §IX, we discuss primitive and transitive transformation groups, and prove the following :

Suppose we have a closed subgroup G of $\mathcal{D}(M)$. Denote by $\mathfrak{g}$ the set of all infinitesimal transformations contained in G, that is, $\mathfrak{g}$ is the set of C^∞-vector fields u on M such that the one parameter subgroup $\exp tu$ generated by u is contained in G. Then, $\mathfrak{g}$ is a closed linear subspace of $\Gamma(T_M)$ (C^∞-vector fields with C^∞-topology) and a Lie subalgebra of $\Gamma(T_M)$. (Cf. § I.)

Theorem H (1) If $\dim \mathfrak{g} < \infty$, then G is a finite dimensional Lie group under the LPSAC-topology.

(2) Suppose $\dim \mathfrak{g} = \infty$. If G satisfies the following conditions (a), (b) and (c), then G has a strong ILH-Lie group structure under the relative topology in $\mathcal{D}(M)$.

(a) G acts transitively on M.

(b) G acts primitively on M, that is, G leaves no foliation invariant and the isotropy subalgebra $\mathfrak{g}_o(x) = \{ u \in \mathfrak{g} : u(x) = 0 \}$ at a point $x \in M$ is a prope maximal subalgebra of $\mathfrak{g}$.

(c) Any $u \in \Gamma(T_M)$ which contacts $\mathfrak{g}$ is contained in $\mathfrak{g}$, where u is called t contact $\mathfrak{g}$ if for any integer r and for any point $x \in M$, there is $v_x \in \mathfrak{g}$ such that $(j^r u)(x) = (j^r v_x)(x)$.

The definition of LPSAC-topology will be given in §I. Here we mention only that this is stronger than the relative topology and makes G locally piecewise-smooth-arcwise connected.

If we assume, instead of the condition (b), that G acts irreducibly on M, i.e. that the linear isotropy of G is irreducible, then the main part of the above theorem has been already proved in [32]. Furthermore the basic idea is not changed in this article. So, one may read [32] for a further introduction. However, this article is selfcontained, we restate the results in the sections where they are proved. Many results in [32] are sharpened in this article.

The proof of Theorem H is separated roughly into four steps. Forst, if $\mathfrak{g}$ is of finite dimension, then we need no other condition, G is always a finite dimensional Lie group under the LPSAC-topology. Second, if $\mathfrak{g}$ is of infinite dimension and satisfies the conditions (a) - (c), then such Lie algebras are classifield [15, 23, 40]. There are the following four algebras : (1) $\Gamma(T_M)$, (2) the Lie algebra of infinitesimal volume preserving transformations, (3) the Lie algebra of infinitesimal symplectic transformations, (4) the Lie algebra of infinitesimal contact transformations.

All algebras except (4) are irreducible. So, taking [32] into account, we may restrict our attention to the contact transformations. However, the above four cases are treated simultaneously.

Third, let $\mathfrak{g}$ be one of the above Lie algebras. Then, by using smooth extension theorems [31], we can prove the smoothness of the right invariant distribution $\tilde{\mathfrak{g}} = dR_g \mathfrak{g}$ on $\mathfrak{D}(M)$ or on some modification of it. Fourth, we apply the Frobenius theorem obtained above.

In §X, we discuss some specific properties of Lie algebras of vector fields. where the theorem of Pursell and Shanks will be extended to Lie algebras of infinitesimal automorphisms of various structures on manifolds. (Cf. [39]) The theorem obtained here includes the following :

The strong ILH-Lie group structure of infinite dimensional, primitive and transitive groups satisfying the conditions of Theorem H is completely determined by its Lie algebra.

The proof of this, however, comes from another direction (not from the theory of strong ILH-Lie groups). We use an algebraic method.

In § XI, we discuss linear groups which contain all groups of diffeomorphisms of compact manifolds.

In this article, the author devotes his attention to contact transformation groups. First, as it is mentioned above, the irreducible case has been already discussed in [32]. Second, it seems interesting to investigate the relation between contact transformation groups and strictly contact transformation groups.

Let M be of odd dimension and assume that M has a C^∞-contact 1-form ω. The contact transformation group $\mathcal{D}_\omega(M)$ is the totality of C^∞-diffeomorphisms ψ of M such that $\psi^*\omega = \tau\omega$ for some function τ and the strictly contact transformation group $\mathcal{D}_{s\omega}(M)$ is defined by $\{ \psi \in \mathcal{D}(M) : \psi^*\omega = \omega \}$. According to Theorem H, $\mathcal{D}_\omega(M)$ is a strong ILH-Lie group.

We call ω a regular contact form, if the characteristic vector field ξ_ω of ω generates a free action of the circle group S^1. Together with the results in § II, we have

Theorem I If ω is a regular contact form, then $\mathcal{D}_{s\omega}(M)$ is a strong ILH-Lie group, and a closed strong ILH-Lie subgroup of $\mathcal{D}_\omega(M)$.

Let ω be a regular contact form on M. Denote by $\mathcal{T}$ the totality of C^∞-functions τ such that $\tau\omega$ are regular contact forms.

Theorem J $\mathcal{T}$ is a closed subset of $\Gamma_*(1_M)$, where $\Gamma_*(1_M)$ is the set of positive or negative functions on M. Moreover, the factor set $\mathcal{D}_{s\omega}(M) \backslash \mathcal{D}_\omega(M)$ is homeomorphic to an open and closed subset of $\mathcal{T}$.

As an immediate corollary of this theorem, we see

Corollary Any deformation of a regular contact form among regular contact forms is locally trivial.

Now, there is another group $\mathcal{D}_{\omega+\alpha}(M)$ containing $\mathcal{D}_{s\omega}(M)$, where $\mathcal{D}_{\omega+\alpha}(M) = \{ \psi \in \mathcal{D}(M) : \psi^*\omega = \omega + \text{closed form} \}$.

Theorem K $\mathcal{D}_{\omega+\alpha}(M)$ is a strong ILH-Lie group, and $\mathcal{D}_{s\omega}(M) \backslash \mathcal{D}_{\omega+\alpha}(M)_o$ is a contractible space and hence $\mathcal{D}_{\omega+\alpha}(M)_o$ is homeomorphic to the direct product of $\mathcal{D}_{s\omega}(M)$ and a contractible space, where the suffix o of $\mathcal{D}_{\omega+\alpha}(M)$ means the subgroup which preserves the orientation of M.

Suggested by the above theorem and Theorem 2.18 in [28], it is natural to conjecture that the factor set $\mathcal{D}_{s\omega}(M) \backslash \mathcal{D}_{\omega}(M)$ is a contractible space, because the only difference are the function factors. However, the situation here is not so simple. The space $\mathcal{T}$ is complicately imbedded in $\Gamma_*(1_M)$ and of course not an open subset of $\Gamma_*(1_M)$. There exists a functional on $\mathcal{D}_{\omega}(M)$ whose extremals form the subgroup $\mathcal{D}_{s\omega}(M)$, but the variational technique does not work without any information about the "shape" of $\mathcal{T}$ in $\Gamma_*(1_M)$.

Theorems I - K will be proved in § VIII.

Anyhow, it is now clear that there are many examples of strong ILB- and strong ILH-Lie groups and that they can be treated somewhat like finite dimensional Lie groups. However, all the concrete examples stated above are not merely strong ILB- (or ILH-) Lie groups. For instance, the Lie algebras of such groups are not only countable Hilbert spaces but also nuclear spaces. In this sense, one may call these groups nuclear groups. Unfortunately, the author could not get any properties of nuclear groups.

The specific character of the group of diffeomorphisms or its subgroups does not consist in the property that the Lie algebra is a nuclear space but in the property that the Lie algebra has a finite codimensional subalgebra. For instance, there exists an example of an infinite dimensional Banach Lie group which acts smoothly, transitively, effectively but not primitively on the two dimensional euclidean space. (Cf. [33].)

Obviously, if a nuclear space has a Banach space structure, then it must be a finite dimensional space.

Although all Hilbert Lie groups are strong ILH-Lie groups and all Banach Lie groups are strong ILB-Lie groups, Banach Lie groups hardly act on finite dimensional manifolds as mentioned on page III. Compairing the statements (a) and (b) in page III. with Theorem H we have to conclude that in order to study groups of diffeomorphisms and their subgroups, we must deal with strong ILH- or strong ILB-Lie groups, however difficult it might be.

Finally, the author wishes to acknowledge to Professor Klingenberg's constant encouragement.

* * * * * * * * * * *

This work was done under the program "Sonderforschungsbereich Theoretische Mathematik (SFB 40)" at the University of Bonn.

Contents

§ I General theory of strong ILB-Lie groups and subgroups

1.1 Definition of strong ILB- (ILH-) Lie groups.

Let $N(d)$ be the set of all integers m such that $m \geq d$. We call a system $\{ \mathbb{E}, E^k, k \in N(d)\}$ a Sobolev chain, if every E^k is a Banach (resp. Hilbert) space, E^{k+1} is linearly and densely imbeded in E^k and $\mathbb{E}$ is the intersection of all E^k with the inverse limit topology. Remark that we do not assume the compactness of the inclusion $E^{k+1} \subset E^k$.

A topological group G is called a strong ILB-(resp. ILH-) Lie group modeled on $\{ \mathbb{E}, E^k, k \in N(d)\}$ (or G has a strong ILB- (resp. ILH-) Lie group structure modeled on $\{ \mathbb{E}, E^k, k \in N(d)\}$), if the following conditions (N,1) - (N,7) are satisfied :

(N,1) There is a Sobolev chain $\{ \mathbb{E}, E^k, k \in N(d)\}$, where every E^k is a Banach (resp. Hilbert) space. There exist an open neighborhood U of 0 in E^d and a homeomorphism ξ of $U \cap \mathbb{E}$ with the relative topology in $\mathbb{E}$ onto an open neighborhood $\tilde{U}$ of e in G, such that $\xi(0) = e$.

(N,2) There exists an open neighborhood V of 0 of E^d such that $\xi(V \cap \mathbb{E}) = \xi(V \cap \mathbb{E})^{-1}$ and $\xi(V \cap \mathbb{E})^2 \subset \xi(U \cap \mathbb{E})$.

(N,3) Letting $\eta(u,v) = \xi^{-1}(\xi(u)\xi(v))$, η can be extended to a C^ℓ-mapping of $V \cap E^{k+\ell} \times V \cap E^k$ into $U \cap E^k$ for every $k \in N(d)$, $\ell \geq 0$.

(N,4) Letting $\eta_v(u) = \eta(u,v)$, η_v can be extended to a C^∞-mapping of $V \cap E^k$ into $U \cap E^k$ for every $v \in V \cap E^k$ and $k \in N(d)$.

(N,5) Letting $\theta(w,u,v) = (d\eta_v)_u w$, θ can be extended to a C^ℓ-mapping of $E^{k+\ell} \times V \cap E^{k+\ell} \times V \cap E^k$ into E^k for every $k \in N(d)$, $\ell \geq 0$.

(N,6) The mapping $\iota : V \cap \mathbb{E} \mapsto V \cap \mathbb{E}$ defined by $\iota(u) = \xi^{-1}(\xi(u)^{-1})$ can be extended to a C^ℓ-mapping of $V \cap E^{k+\ell}$ into $V \cap E^k$ for every $k \in N(d)$, $\ell \geq 0$.

(N,7) For any $g \in G$, there exists an open neighborhood W of 0 of E^d such that

$g^{-1}\xi(W \cap \mathbb{E})g \subset \xi(U \cap \mathbb{E})$, and the mapping A_g defined by $A_g(u) = \xi^{-1}(g^{-1}\xi(u)g)$ can be extended to a smooth mapping of $W \cap E^k$ into $U \cap E^k$ for every $k \in N(d)$.

1.2 Global properties of strong ILB-(ILH-) Lie groups.

The goal of this section is the proof of the following :

1.2.1 Theorem <u>A topological group</u> G <u>is a strong</u> ILB-(resp. ILH-) <u>Lie group modeled on</u> $\{\mathbb{E}, E^k, k \in N(d)\}$, <u>if and only if there exists a system</u> $\{G^s, s \in N(d)\}$ <u>of topological groups</u> G^s <u>which satisfy the conditions</u> (G,1) - (G,8) <u>stated below</u>.

(G,1) G^s is a C^∞-Banach (resp. Hilbert) manifold modeled on E^s and a topological group.

(G,2) G^{s+1} is a dense subgroup of G^s and the inclusion is of class C^∞.

(G,3) $G = \cap G^s$ with the inverse limit topology and with the inverse limit of group structures.

(G,4) The group multiplication $G \times G \mapsto G$, $(g,h) \mapsto gh$, can be extended to the C^ℓ-mapping of $G^{s+\ell} \times G^s$ onto G^s.

(G,5) The mapping $G \mapsto G$, $g \mapsto g^{-1}$, can be extended to the C^ℓ-mapping of $G^{s+\ell}$ into G^s.

(G,6) For any $g \in G^s$, the right translation $R_g : G^s \mapsto G^s$ is of class C^∞.

(G,7) Let $\mathfrak{g}^s$ be the tangent space of G^s at the identity e, and let T_{G^s} be the tangent bundle of G^s. The mapping $dR : \mathfrak{g}^{s+\ell} \times G^s \mapsto T_{G^s}$, defined by $dR(u,g) = dR_g u$, is of class C^ℓ.

(G,8) There are open neighborhood U of 0 in $\mathfrak{g}^d$ and a C^∞-diffeomorphism ξ of U onto an open neighborhood $\tilde{U}$ of e in G^d with $\xi(0) = e$ such that the restriction of ξ onto $U \cap \mathfrak{g}^s$ gives a C^∞-diffeomorphism of the open subset $U \cap \mathfrak{g}^s$ in $\mathfrak{g}^s$ onto the open subset $\tilde{U} \cap G^s$ in G^s for any $s \geq d$.

Before we start to prove the above theorem, we will give the following definition and **remarks**.

1.2.2 Definition Let G, H be strong ILB-(or ILH-) Lie groups modeled on $\{\mathbb{E}, E^k, k \in N(d)\}$, $\{\mathbb{F}, F^k, k \in N(d')\}$ respectively. We call that G and H are equivalent, if there is an isomorphism i of G onto H such that i can be extended to a C^∞-isomorphism of G^s onto H^s for all $s \geq d''$, where d" is an integer such that $d'' \geq \max(d, d')$.

Remark 1. Roughly speaking, the above condition (G,8) implies that a neighborhood of the identity of every G^s can be coordinatized by a mapping which is independent from s. So, letting $\mathfrak{g} = \cap \mathfrak{g}^s$ with the inverse limit topology, we see immediately that $\xi : U \cap \mathfrak{g} \mapsto \tilde{U} \cap G$ is a homeomorphism and gives a Frechet Lie group structure on G.

The pair (U, ξ) in the condition (G,8) will be called an ILB-(or ILH-) coordinate of G at the identity.

Remark 2. A topological group G is called an ILB-(or ILH-) Lie group, if there is a system $\{G^s, s \in N(d)\}$ satisfying the conditions (G,1) - (G,7) above. There are several examples of ILH-Lie groups. In their interesting paper [10], Ebin and Marsden showed that the following groups are ILH-Lie groups :

(a) The group of diffeomorphisms which leaves a volume element invariant.

(b) The group of diffeomorphisms which leaves a symplectic structure invariant.

The above results were obtained by using the implicit function theorem on Hilbert manifolds. Furthermore, the author gave other examples of ILH-Lie groups by using Frobenius theorem on Hilbert manifolds. (See [31] pp415-420.) These results are obtained roughly by lifting up everything on G to the stage G^s, where the implicit function or Frobenius theorem can be applied. Thus, the coordinate neighborhoods used in the proofs may depend on s, hence the resulting group might not be a Frechet manifold.

This is a week point of the above results, as the author pointed out already in the introduction of [32]. Namely, as far as concerning about ILH-Lie groups, the inverse limit $G = \lim G^s$ might not be a Frechet manifold (and probably there is a counter example). On the other hand, the group of C^∞-diffeomorphisms on a closed manifold is not only an ILH-Lie group and a Frechet Lie group but also has some nicer properties mentioned above. (Cf. §II)

Proof of 1.2.1.

It is easy to see that if G satisfies (G,1) - (G,8), then G is a strong ILB- (resp. ILH-) Lie group. So, we have only to show the converse.

Notations being as in (N,1) - (N,7) above, there exists a neighborhood V_1 of 0 in E^d such that $\eta(\iota(V_1), V_1) \subset V$ (cf. (N,3-4)), hence $\eta(\iota(V_1 \cap \mathbb{E}), V_1 \cap \mathbb{E}) \subset V \cap \mathbb{E}$, that is, $\xi(V_1 \cap \mathbb{E})^{-1}\xi(V_1 \cap \mathbb{E}) \subset \xi(V \cap \mathbb{E})$.

We put $\tilde{V}_1 = \xi(V_1 \cap \mathbb{E})$. Then, $\tilde{V}_1 \cdot g$ is an open neighborhood of g in G for every $g \in G$. Let $\xi_g : V_1 \cap \mathbb{E} \mapsto \tilde{V}_1 \cdot g$ be the homeomorphism defined by $\xi_g(u) = \xi(u)g$. If $\tilde{V}_1 g \cap \tilde{V}_1 h$ contains an element g', then $hg^{-1} = hg'^{-1}g'g^{-1} \in \tilde{V}_1{}^{-1}\tilde{V}_1 \subset \xi(V \cap \mathbb{E})$. We put $\xi(u) = g'g^{-1}$, $\xi(v) = g'h^{-1}$, $u, v \in V_1 \cap \mathbb{E}$ and $\xi(w) = hg^{-1}$, $w \in V \cap \mathbb{E}$. Then, $\xi(v)\xi(w) = \xi(u)$, i.e. $\eta(v,w) = u$. Thus, there exists the maximal open subset $W'(g')$ in V_1 such that $W'(g') \ni v$ and $\eta(W'(g'),w) \subset V_1$. We put $W(g') = \eta(W'(g'),w)$. Thus, the mapping $\xi_g^{-1}\xi_h : W'(g') \cap \mathbb{E} \mapsto W(g') \cap \mathbb{E}$ can be extended to a smooth diffeomorphism of $W'(g') \cap E^k$ onto $W(g') \cap E^k$ for every $k \geq d$, because $\xi_g^{-1}\xi_h(v') = \eta(v',w)$. (Cf. (N,4).)

We put

$$U_{h,g} = \cup \{W'(g') \; ; \; g' \in \tilde{V}_1 g \cap \tilde{V}_1 h \},$$

$$U_{g,h} = \cup \{W(g') \; ; \; g' \in \tilde{V}_1 g \cap \tilde{V}_1 h \}.$$

$U_{h,g}$ and $U_{g,h}$ are open subset of V_1, and obviously $\xi_h(U_{h,g} \cap \mathbb{E}) = \xi_g(U_{g,h} \cap \mathbb{E}) = \tilde{V}_1 g \cap \tilde{V}_1 h$.

The following is easy to prove :

1.2.3 Lemma <u>The mapping</u> $\xi_g^{-1}\xi_h : U_{h,g} \cap \mathbb{E} \mapsto U_{g,h} \cap \mathbb{E}$ <u>can be extended to a smooth</u>

diffeomorphism of $U_{h,g} \cap E^k$ onto $U_{g,h} \cap E^k$ for every $k \geq d$. Therefore, the maximal open subset $W'(g')$, $W(g')$ are in fact equal to $U_{h,g}$, $U_{g,h}$ respectively.

The pair $(V_1 \cap \mathbb{E}, \xi_g)$ can be regarded as a chart of G around g. So, we consider ξ_g as an index and make an atlas $\mathfrak{A}$ of G by the disjoint union $\cup \{(V_1 \cap \mathbb{E}), \xi_g) : g \in G\}$. The space G is then naturally identified with the union of the charts glued up by the equivalence relation $\sim$, where the equivalence relation is defined as follows : $u \in (V_1 \cap \mathbb{E}, \xi_g)$ and $v \in (V_1 \cap \mathbb{E}, \xi_h)$ is equivalent, if and only if $\xi_g(u) = \xi_h(v)$.

Now, consider the disjoint union $\cup \{(V_1 \cap E^k, \xi_g) : g \in G\}$ and regard ξ_g as an index. We define an equivalence relation $\sim$ by the following :

For elements $u \in (V_1 \cap E^k, \xi_g)$ and $v \in (V_1 \cap E^k, \xi_h)$, u is equivalent with v (notation $u \sim v$), if and only if $u \in U_{g,h} \cap E^k$, $v \in U_{h,g} \cap E^k$ and $\xi_g^{-1}\xi_h(v) = u$, where $\xi_g^{-1}\xi_h$ is the extended diffeomorphism. (Cf. 1.2.3.) (It is not hard to verify that the above $\sim$ is an equivalence relation.)

Let G^k be the union of $(V_1 \cap E^k, \xi_g)$, $g \in G$, glued up by that equivalence relation. Then, G^k is obviously a smooth Banach (resp. Hilbert) manifold modeled on E^k. $(V_1 \cap E^k, \xi_g)$ is a chart of G^k around g. We denote by ξ_g the natural injection of $V_1 \cap E^k$ into G^k. Thus, we have a system $\{G, G^k, k \in N(d)\}$ satisfying the following :

(M,1) Every G^k is a smooth Banach (resp. Hilbert) manifold modeled on E^k.

(M,2) $G^{k+1} \subset G^k$ and the inclusion is smooth.

(M,3) $G = \cap G^k$ with the inverse limit topology and G is dense in every G^k.

(M,4) $(V_1 \cap E^k, \xi_g)$ is a smooth chart of G^k for any $g \in G$ and satisfies
$$\xi_g(V_1 \cap E^k) = \xi_g(V_1) \cap G^k.$$

For the proof of 1.2.1, we have to show at first that G^k is a topological group. This will be done in the following three lemmas.

Let U be the same neighborhood as in (N,1). Let $\mathfrak{N}^k$ be a basis of neighborhoods of 0 in E^k such that every $W \in \mathfrak{N}^k$ is contained in $U \cap E^k$.

1.2.4 Lemma <u>The system of neighborhoods</u> $\{ \xi(W \cap \mathbb{E}) : W \in \mathfrak{N}^k \}$ <u>defines a weaker topology for</u> G <u>than the original one. By this topology,</u> G <u>becomes a topological group.</u>

Proof. We have only to show that the system $\{ \xi(W \cap \mathbb{E}) : W \in \mathfrak{N}^k \}$ satisfies the axioms of neighborhoods of the identity of topological groups.

(a) $\cap \{\xi(W \cap \mathbb{E}) : W \in \mathfrak{N}^k \} = \{ e \}$. (Trivial.)

(b) For any $W_1, W_2 \in \mathfrak{N}^k$, there is W_3 such that $\xi(W_1 \cap \mathbb{E}) \cap \xi(W_2 \cap \mathbb{E}) \supset \xi(W_3 \cap \mathbb{E})$. (Also, trivial.)

(c) For any $W_1 \in \mathfrak{N}^k$, there is $W_2 \in \mathfrak{N}^k$ such that $\xi(W_2 \cap \mathbb{E})\xi(W_2 \cap \mathbb{E})^{-1} \subset \xi(W_1 \cap \mathbb{E})$. (Easy to prove by (N,3) and (N,6).)

(d) For any $W_1 \in \mathfrak{N}^k$ and for any element $g = \xi(u)$, $u \in W_1 \cap \mathbb{E}$, there is $W_2 \in \mathfrak{N}^k$ such that $\xi(W_2 \cap \mathbb{E})\xi(u) \subset \xi(W_1 \cap \mathbb{E})$. (Cf. (N,3))

(e) For any $W_1 \in \mathfrak{N}^k$ and for any $g \in G$, there is $W_2 \in \mathfrak{N}^k$ such that $g^{-1}\xi(W_2 \cap \mathbb{E})g \subset \xi(W_1 \cap \mathbb{E})$. (Cf. (N,7).)

Let $\overline{G}^k$ denote the completion of G by the right uniform topology defined by the above new topology $\{ \xi(W \cap \mathbb{E}) : W \in \mathfrak{N}^k \}$. Obviously, $\overline{G}^k$ is a topological group.

1.2.5 Lemma <u>Notations are as above. For every</u> $v, w \in V \cap E^k$, <u>there are neighborhoods</u> W_v, W_w <u>of</u> v, w <u>in</u> $V \cap E^k$ <u>and a positive constant</u> C <u>such that</u>

$$\| \theta(u, v', w') \|_k \leq C \, \|u\|_k$$

<u>for any</u> $v' \in W_v$ <u>and</u> $w' \in W_w$ <u>, where</u> $\| \ \|_k$ <u>is the norm on</u> E^k.

Proof. By the property (N,5), the mapping $\theta : E^k \times V \cap E^k \times V \cap E^k \mapsto E^k$ is continuous. Since $\theta(0,v,w) = 0$, there are a δ-neighborhood $W(\delta)$ of 0 of E^k and neighborhoods W_v, W_w of v, w respectively such that $\|\theta(u',v',w')\|_k \leq 1$ for any $u' \in W(\delta)$, $v' \in W_v$, $w' \in W_w$. Since θ is linear with respect to the first variable letting $C = 1/\delta$, we get the desired result.

1.2.6 Lemma $G^k = \bar{G}^k$

Proof. At first, we prove $\bar{G}^k \subset G^k$. Let $\{g_n\}$ be a Cauchy sequence in G in the right uniform topology defined above. Then, $\xi^{-1}(g_n g_m^{-1})$ can be defined for sufficiently large n, m and $\|\xi^{-1}(g_n g_m^{-1})\|_k \mapsto 0$ as $n, m \mapsto \infty$. By the above lemma, there is a closed and convex neighborhood W of 0 in $V \cap E^k$ such that $\|\theta(u,v,w)\|_k \leq C\|u\|_k$ for every $v, w \in W$. By the properties (N,3) and (N,6), there is a closed convex neighborhood W_1 of 0 in $V \cap E^k$ such that $\xi(W_1 \cap \mathbb{E})^{-1} \subset \xi(W \cap \mathbb{E})$ and $\xi(W_1 \cap \mathbb{E})^2 \subset \xi(W \cap \mathbb{E})$.

Since $\{g_n\}$ is a Cauchy sequence, there is a sufficiently large number N such that for every $n \geq N$, $g_n g_N^{-1}$ is contained in $\xi(W_1 \cap \mathbb{E})$. Put $u_n = \xi^{-1}(g_n g_N^{-1})$. Then, we have

$$u_n - u_m = \eta(\eta(u_n, \iota(u_m)), u_m) - \eta(\eta(u_m, \iota(u_m)), u_m)).$$

Since $\eta(u_m, \iota(u_m)) = 0$, we have

$$u_n - u_m = \int_0^1 \theta(\eta(u_n, \iota(u_m)), t\eta(u_n, \iota(u_m)), u_m)\, dt .$$

On the other hand, $\xi\eta(u_n, \iota(u_m)) = g_n g_N^{-1}(g_m g_N^{-1})^{-1} \in \xi(W \cap \mathbb{E})$. Thus, we have $\|u_n - u_m\|_k \leq C\|\eta(u_n, \iota(u_m))\|_k \longmapsto 0$ as $n, m \mapsto \infty$. Therefore, $\{u_n\}$ is a Cauchy sequence in $W_1 \cap E^k$, hence converges to an element $u_o \in W_1$. Thus, $g_n = \xi(u_n)g_N$ converges to $\xi(u_o)g_N$, which is obviously contained in G^k.

Conversely, suppose we have a sequence $\{g_n\}$ in G converging to an element g_o in G^k. To prove $G^k \subset \bar{G}^k$, we have only to show that $\{g_n\}$ is a Cauchy sequence in the right uniform topology. For the proof, we may assume that there is $h \in G$ such that g_o and all g_n are contained in $\xi(V_1 \cap E^k)h$. Put $g_n = \xi(u_n)h$, $g_o = \xi(u_o)h$. We see easily that $\{u_n\}$ converges to u_o in $V_1 \cap E^k$. Thus,

$$\xi^{-1}(g_n g_m{}^{-1}) = \xi^{-1}(\xi(u_n)\xi(u_m)^{-1}) = \eta(u_n, \iota(u_m))$$

$$= \int_0^1 \theta(u_n - u_m, u_m + t(u_n - u_m), \iota(u_m))dt.$$

Since $\lim u_n = u_o$, $\lim \iota(u_n) = \iota(u_o)$ and $u_o \in V_1 \cap E^k \subset U \cap E^k$, we see easily by the above lemma that

$$\|\theta(u_n - u_m, u_m + t(u_n - u_m), \iota(u_m))\|_k \leq C\|u_n - u_m\|_k$$

for sufficiently large n, m. Hence $\|\xi^{-1}(g_n g_m{}^{-1})\|_k \mapsto 0$ as $n, m \mapsto \infty$. This implies that $\{g_n\}$ is a Cauchy sequence in the right uniform topology.

By the above three lemmas, we see that the system $\{ G, G^k, k \in N(d)\}$ satisfies (G,1), (G,2), (G,3) and (G,8). However, all other properties (G,4 ~ 7) are the reflections of the local properties (N,3 ~ 7), and the proof is not hard. Thus, 1.2.1 is proved.

Now, in the last part of this section, we remark the following :

1.2.7 Proposition <u>The inclusion</u> $i : G \mapsto G^k$ <u>induces an isomorphism of homotopy groups</u> $\pi_*(G)$ <u>and</u> $\pi_*(G^k)$ <u>for any</u> $k \in N(d)$.

If G satisfies the second countability axiom, then by Theorem 15 [37], the above inclusion gives a homotopy equivalence.

The above proposition is an immediate conclusion of the following approximation lemma :

1.2.8 Lemma <u>Let</u> $\{ \mathbb{E}, E^k, k \in N(d)\}$ <u>be a Sobolev chain and</u> U <u>be a convex open subset of</u> E^d. <u>Suppose</u> $f : D^m \mapsto U \cap E^k$ <u>is a continuous mapping of an m-disk</u> D^m <u>into</u> $U \cap E^k$ <u>such that</u> $f|\partial D^m$ <u>is a composition of the inclusion</u> $i : U \cap \mathbb{E} \mapsto U \cap E^k$ <u>and a continuous mapping</u> g <u>of</u> $S^{m-1} = \partial D^m$ <u>into</u> $U \cap \mathbb{E}$. <u>Then, there is a homotopy</u> $F : D^m \times [0,1] \mapsto U \cap E^k$ <u>such that</u> $F(*,0) = f$, $F(x,t) = ig(x)$ <u>for every</u> $x \in S^{m-1}$ <u>and</u> $F(*,1)$ <u>is a composition</u> iG <u>of</u> i <u>and a continuous mapping</u> G <u>of</u> D^m <u>into</u> $U \cap \mathbb{E}$.

Proof. Let 0 be the center of D^m. Choose an arbitrary point in $U \cap \mathbb{E}$ and denote this by $G(0)$. Every point of D^m is described by $s\bar{x}$, $s \in [0,1]$, $\bar{x} \in S^{m-1}$. Then, the desired homotopy F is given by

$$F(s\bar{x},t) = (1 - t)f(s\bar{x}) + t(s\, i\cdot g(\bar{x}) + (1 - s)G(0)).$$

I.3 Lie algebras, exponential mappings and Lie algebra homomorphisms.

Let $\mathfrak{g}^k$ be the tangent space of G^k at the identity e and let $\mathfrak{g} = \cap\, \mathfrak{g}^k$ with the inverse limit topology. $\mathfrak{g}^k$ is obviously isomorphic to E^k and so is $\mathfrak{g}$ to $\mathbb{E}$. Now, we define a Lie algebra structure on $\mathfrak{g}$ by the following manner : For any $u, v \in \mathfrak{g}$, we put $\tilde{u}(g) = dR_g u$, $\tilde{v}(g) = dR_g v$ for any $g \in G^k$. By the property (G,7), we see that $\tilde{u}, \tilde{v}$ are C^∞-vector fields on every G^k. Define $[u, v] = (\widetilde{uv} - \widetilde{vu})(e)$. This definition does not depend on k and hence gives a Lie algebra structure on $\mathfrak{g}$. The mapping $[\ ,\ ] : \mathfrak{g} \times \mathfrak{g} \mapsto \mathfrak{g}$ can be extended to the bilinear, bounded operator of $\mathfrak{g}^{k+1} \times \mathfrak{g}^{k+1}$ into $\mathfrak{g}^k$. We call $\mathfrak{g}$ <u>the Lie algebra of</u> G.

Now, for every element $u \in \mathfrak{g}^{k+\ell}$, we denote by $\tilde{u}$ the vector field of G^k defined by $\tilde{u}(g) = dR_g u$. $\tilde{u}$ is a C^ℓ-vector field on G^k.

Suppose $\ell \geq 1$. Then, we can define integral curves of $\tilde{u}$, and since $\tilde{u}$ is right-invariant, we see that $\tilde{u}$ is complete. Let $\exp tu$ be the integral curve through the identity.

1.3.1 Lemma <u>Suppose</u> $\ell \geq 1$. <u>Then, the mapping</u> $u \mapsto \exp u$ <u>is a</u> C^ℓ<u>-mapping of</u> $\mathfrak{g}^{k+\ell}$ <u>into</u> G^k.

Proof. $\mathfrak{g}^{k+\ell} \times G^k$ is a smooth Banach manifold and the mapping $(u,g) \mapsto (0, dR_g u)$ defines a C^ℓ-vector field $\tilde{X}$ on $\mathfrak{g}^{k+\ell} \times G^k$. Let $c(t)$ be an integral curve of $\tilde{X}$ with the initial condition (u,e). Then, $c(t)$ is contained in $\{u\} \times G^k$ and $\pi c(t) = \exp tu$, where $\pi : \mathfrak{g}^{k+\ell} \times G^k \mapsto G^k$ is the projection. So, the differentiability of c with respect to the initial condition yields the desired result.

<u>Remark</u>. It is very likely that $\exp : \mathfrak{g} \mapsto G$, which is defined by the inverse limit, can be extended to the continuous mapping of $\mathfrak{g}^k$ into G^k. However, this is not necessarily a C^1-mapping, hence we can not use the implicit function theorem. Moreover, there is an example such that the exponential mapping does not cover a neighborhood of e. (Cf. the introduction of [28].)

Of course, we call the mapping $\exp : \mathfrak{g} \mapsto G$ <u>the exponential mapping.</u>

Let G, H be strong ILB-Lie groups modeled on Sobolev chains $\{\mathbb{E}, \mathbb{E}^k, k \in N(d)\}$, $\{\mathbb{F}, \mathbb{F}^k, k \in N(d)\}$ respectively. Let $\mathfrak{g}$, $\mathfrak{h}$ be Lie algebras of G, H respectively. Suppose we have a continuous homomorphism φ of $\mathfrak{g}$ into $\mathfrak{h}$. Then, by the definition of continuity, we see that for every $\mathfrak{h}^k$, there is $\mathfrak{g}^{j(k)}$ such that the mapping $\varphi : \mathfrak{g} \mapsto \mathfrak{h}$ can be extended to a bounded linear operator of $\mathfrak{g}^{j(k)}$ into $\mathfrak{h}^k$. Assume furthermore that G is connected and simply connected. Then, we have

1.3.2 Theorem <u>Notations and assumptions being as above, there exists a homomorphism</u> $\Phi : G \mapsto H$ <u>such that</u> Φ <u>can be extended to a</u> $C^{\ell-1}$<u>-mapping of</u> $G^{j(k)+\ell}$ <u>into</u> $H^{k-\ell}$ <u>for every</u> $k > \ell + 1, k \in N(d)$ <u>and the derivative</u> $(d\Phi)_e$ <u>of</u> Φ <u>at the identity is equal to the extended homomorphism</u> $\varphi : \mathfrak{g}^{j(k)+\ell} \mapsto \mathfrak{h}^{k-\ell}$.

Proof. Consider the direct product $G^{j(k)+\ell} \times H^{k-\ell}$. This is a smooth Banach manifold. Let π be the projection of $G^{j(k)+\ell} \times H^{k-\ell}$ onto $H^{k-\ell}$. Let T, T' be the tangent bundle of $G^{j(k)+\ell} \times H^{k-\ell}$ and the pull back of the tangent bundle of $H^{k-\ell}$ by the projection π.

Let (u,v) be a tangent vector at $(g,h) \in G^{j(k)+\ell} \times H^{k-\ell}$. We define a mapping $\tilde{A} : T \mapsto T'$ by $\tilde{A}(u,v) = v - dR_h \varphi\, dR_{g^{-1}} u$. Obviously, this is a right-invariant mapping which preserves the fibres.

Let i be the inclusion map of $G^{j(k)+\ell}$ into $G^{j(k)}$. Since $u \in \mathfrak{g}^{j(k)+\ell}$ and $dR_h \varphi\, dR_{g^{-1}} u = dR_h \varphi\, dR_{i(g)^{-1}} u$, we see by the property (G,7) that $\tilde{A}$ is a C^ℓ mapping, and hence $\tilde{A}$ is a $C^{\ell-1}$-bundle morphism. (Cf. Remark 2 below.) $\tilde{A}$ is obviously a surjection. Therefore, the kernel of $\tilde{A}$ is a $C^{\ell-1}$-subbundle of T, that is, a $C^{\ell-1}$-distribution of $G^{j(k)+\ell} \times H^{k-\ell}$. Since $\tilde{A}$ is invariant under the right translations, the kernel of $\tilde{A}$ is a right-invariant distribution. This distribution will be denoted by $\tilde{\gamma}^k$.

Let γ be the subspace given by $\{(u, \varphi(u)) : u \in \mathfrak{g}\}$. Since φ is a homomorphism, γ is a Lie subalgebra of $\mathfrak{g} \times \mathfrak{h}$. Since $\tilde{\gamma}^k$ contains $dR_{(g,h)}\gamma$ for every $(g,h) \in G^{j(k)+\ell} \times H^{k-\ell}$, we see that $\tilde{\gamma}^k$ is a $C^{\ell-1}$-right invariant involutive distribution of $G^{j(k)+\ell} \times H^{k-\ell}$. (Cf. Proposition A of [31].) Therefore, by Frobenius theorem of Banach manifolds, we have a maximal integral submanifold Γ^k

through the identity. Since the distribution is right-invariant, we see that Γ^k is a subgroup of $G^{j(k)+\ell} \times H^{k-\ell}$. Γ^k is a $C^{\ell-1}$-submanifold and gives a graph of a $C^{\ell-1}$ homomorphism $\Phi^k : G^{j(k)+\ell} \mapsto H^{k-\ell}$, because $G^{j(k)+\ell}$ is connected and simply connected. (Cf. 1.2.7) The derivative $d\Phi^k$ of Φ^k at e is obviously equal to φ.

Since Γ^k is given as the maximal integral submanifold through the identity, we have that $\Gamma^{k+1} \subset \Gamma^k$ and the inclusion is of class $C^{\ell-1}$. Thus, the homomorphism Φ^k is an extension of Φ^{k+1}. In this sense, we omit the suffix k of Φ. The mapping Φ, in fact, this is a system of mappings Φ^k, induces a continuous homomorphism Φ of G into H. This is the desired homomorphism.

1.3.3 Corollary <u>Natations and assumptions being as above, assume furthermore that H is connected and simply connected and $\mathfrak{g}$ is isomorphic to $\mathfrak{h}$. Then, G is isomorphic to H.</u>

<u>Remark 1</u> If G is a strong ILB-Lie group, then the universal covering group $\tilde{G}$ is also a strong ILB-Lie group. So, the above Corollary shows that the group structures of strong ILB-Lie groups are locally determined by Lie algebras. However, it seems to be very hard to say about strong ILB-Lie group structures from the structures of Lie algebras.

<u>Remark 2</u> Let $A : B_1 \mapsto B_2$ be a fibre preserving mapping of a vector bundle B_1 over a Banach manifold X with a fibre E_1 into another vector bundle B_2 over X with a fibre E_2. Suppose E_1, E_2 are Banach spaces. Since B_1, B_2 are Banach manifolds, we can define the concept that A is a C^k-mapping. However, it is slightly different from the concept of C^k-bundle morphism.

Let U be a smooth chart of X. Then, the mapping A induces a fibre preserving mapping $A' : U \times E_1 \mapsto U \times E_2$. We put $A'(x,v) = (x, A'_x v)$. Now, if A is a C^k-mapping, then this implies that the mapping $A' : U \times E_1 \mapsto U \times E_2$ is a C^k-mapping, but when we say A is a C^k-bundle morphism, this means that the mapping $A'' : U \mapsto L(E_1,E_2)$ defined by $x \mapsto A'_x$ is a C^k-mapping , where $L(E_1,E_2)$ is the Banach space of all bounded linear operators of E_1 into E_2 with the operator norm topology.

In general, there is the following relation :

1.3.4 Lemma <u>Let</u> E, F, F' <u>be Banach spaces. Let</u> U <u>be an open subset of</u> E. <u>Assume</u> $A : F\times U \mapsto F'$ <u>is a</u> C^k<u>-mapping and</u> A <u>is linear with respect to the first variable. Then, the mapping</u> $\widetilde{A} : U \mapsto L(F,F')$ <u>defined by</u> $\widetilde{A}(g)v = A(v,g)$ <u>is a</u> C^{k-1}<u>-mapping, where</u> $k \geq 1$. <u>If</u> $k = 0$, <u>then</u> $\widetilde{A}$ <u>is locally bounded, that is, for any</u> $x \in U$, <u>there is a neighborhood</u> V_x <u>of</u> x <u>and a constant</u> C <u>such that</u> $|||\widetilde{A}(g)||| \leq C$.

Proof. For $j \leq k$, the j-th derivative $(d_2^jA)_{(v,g)}$ at (v,g) with respect to the second variable is a bounded, symmetric and j-linear mapping of $E\times\cdots\times E$ into F' and moreover this is also a bounded linear mapping with respect to $v \in F$. We denote by $D^j\widetilde{A}(g)$ the multi-linear mapping of $E\times\cdots\times E\times F$ into F' defined by

$$D^j\widetilde{A}(g)(u_1,\dots,u_j,v) = (d_2^jA)_{(v,g)}(u_1,\dots,u_j).$$

This is continuous with respect to the variables $(g, u_1,\dots, u_j, v)$, because A is a C^k-mapping. Since $D^j\widetilde{A}$ is continuous at $(g,0,\dots,0)$, there is an open neighborhood V of g and a constant C such that

$$\| D^j\widetilde{A}(g') \|_{L^j_{sym}(E,L(F,F'))} \leq C$$

for every $g' \in V$.

We see that

$$\{(D^{j-1}\widetilde{A}(g') - D^{j-1}\widetilde{A}(g)\}(u_1,\dots,u_{j-1},v) = \int_0^1 (d_2^jA)_{g+t(g'-g)}(u_1,\dots,u_{j-1}, g'-g)\,dt$$

Thus,

$$\|D^{j-1}\widetilde{A}(g') - D^{j-1}\widetilde{A}(g)\|_{L^{j-1}_{sym}(E,L(F,F'))} \leq C\int_0^1 \|g'-g\|_E\,dt = C\|g'-g\|_E .$$

Therefore, $D^{j-1}\widetilde{A} : U \mapsto L^{j-1}_{sym}(E,L(F,F'))$ is continuous. Moreover, we see that

$$\widetilde{A}(g') - \widetilde{A}(g) - D\widetilde{A}(g)(g'-g) - \cdots - \frac{1}{(j-1)!}D^{j-1}\widetilde{A}(g)(g'-g)^{j-1}$$

$$= \{\int_0^1 \frac{(1-t)^{j-1}}{(j-1)!}(D^j\widetilde{A})(g+t(g'-g))dt\}(g'-g)^j.$$

Since $\| D^j\widetilde{A}(g')\|_{L^j_{sym}(E,L(F,F')} \leq C$ for every $g' \in V$, we have that

$$\lim_{g'\mapsto g} \frac{1}{\|g'-g\|_E^{j-1}}|||\widetilde{A}(g') - \widetilde{A}(g) - D\widetilde{A}(g)(g'-g) - \cdots - \frac{1}{(j-1)!}D^{j-1}\widetilde{A}(g)(g'-g)^{j-1}|||$$

$= 0,$

where $|||\ |||$ means the operator norm in $L(F,F')$. Therefore, by Theorem 3 in [26 ,p7], we see that $\tilde{A}$ is a C^{k-1}-mapping.

I.4 Subgroups of strong ILB-Lie groups

Suppose we have a topological group H with a basis of neighborhoods $\mathcal{N}$ of the identity e of H. For any $\tilde{U} \in \mathcal{N}$, take the arcwise connected component $\tilde{U}_o$ of $\tilde{U}$ containing e, and make a family $\mathcal{N}_o = \{ \tilde{U}_o : \tilde{U} \in \mathcal{N} \}$. Then, $\mathcal{N}_o$ satisfies the axioms of neighborhoods of e of a topological group. This topology will be called the associated locally arcwise connected topology (briefly, LAC-topology in this article).

Suppose now H is a subgroup with the relative topology of a strong ILB-Lie group G. $\mathcal{N}$ means a basis of neighborhoods of e of G. For any $\tilde{U} \in \mathcal{N}$, let $\tilde{U}_o(H)$ denote the subset consisting of all points x in $\tilde{U} \cap H$ such that x and e can be joined by a piecewise smooth curve in $\tilde{U} \cap H$, where a curve $c(t)$ is called piecewise smooth if the mapping $c : [0,1] \mapsto G^k$ is piecewise of class C^1 for any $k \in N(d)$. Let $\mathcal{N}_o(H) = \{ \tilde{U}_o(H) : \tilde{U} \in \mathcal{N} \}$, and similarly as above, $\mathcal{N}_o(H)$ satisfies the axioms of neighborhoods of e of topological groups. This topology will be called <u>the LPSAC-topology</u> in this article.

It is known that any subgroup H of a finite dimensional Lie group is a Lie group under the LAC-topology. (Cf. [14], 7.4 Corollary.) Another word, any subgroup of a finite dimensional Lie group can be regarded as a Lie subgroup. However, it depends on what definition we pick up as Lie subgroups. In the following, we will fix the definition of strong ILB-Lie subgroups.

Let $\{ \mathbb{E}, E^k, k \in N(d)\}$ be a Sobolev chain and $\mathbb{F}$ a closed subspace of $\mathbb{E}$. Let F^k denote the closure of $\mathbb{F}$ in E^k. Suppose there is a splitting $\mathbb{E} = \mathbb{F} \oplus \mathbb{G}$. Let G^k denote the closure of $\mathbb{G}$ in E^k. The splitting $\mathbb{E} = \mathbb{F} \oplus \mathbb{G}$ is called an <u>ILB-splitting,</u> if $E^k = F^k \oplus G^k$ (direct sum) for every $k \in N(d')$, d' is an integer such that $d' \geq d$. This is called an ILH-normal splitting, if moreover the following inequality holds for every $u \in \mathbb{F}$, $v \in \mathbb{G}$ and $k \geq d'$:

$$\| u + v \|_k \geq C\{\|u\|_k + \|v\|_k\} - D_k\{\|u\|_{k-1} + \|v\|_{k-1}\},$$

where C, D_k are positive constants such that C is independent from k. Since $\|u + v\|_k \leq \|u\|_k + \|v\|_k$, the above inequality means that the norms on E^k are very close to the product norms $\|u\|_k + \|v\|_k$ or $(\|u\|_k^2 + \|v\|_k^2)^{\frac{1}{2}}$.

Let G be a strong ILB-Lie group, $\mathfrak{g}^k$ the tangent space of G^k at e. $\mathfrak{g}$ is the inverse limit of $\mathfrak{g}^k$, $k \in N(d)$. A subgroup H if G is called a strong ILB-Lie subgroup, if the following conditions are satisfied :

(Sub.1) There is an ILB-splitting $\mathfrak{g} = \mathfrak{h} \oplus \mathbb{F}$.
(Sub.2) There is an open neighborhood U' of 0 of $\mathfrak{g}^{d'}$ and regarding G as a strong ILB-Lie group modeled on $\{\mathfrak{g}, \mathfrak{g}^k, k \in N(d')\}$ there is an ILB-coordinate (U', ξ') of G at the identity such that $\xi'(U' \cap \mathfrak{h}) = \tilde{U}'_o(H)$, where $\tilde{U}' = \xi'(U' \cap \mathfrak{g})$.

Obviously, a strong ILB-Lie subgroup is a strong ILB-Lie group.

Now, it is natural to ask to what extent a subgroup with LPSAC-topology of a strong ILB-Lie group is near to a strong ILB-Lie subgroup. This seems, however, a difficult question and is far from the complete settlement. Here, we will give only a first approximation.

1.4.1. Theorem Let G be a strong ILB-Lie group with the Lie algebra $\mathfrak{g}$ and let H be a subgroup of G. Let $\mathfrak{h}$ denote the totality of $u \in \mathfrak{g}$ such that $\exp tu \in H$ for all t. If there is a neighborhood $\tilde{U}$ of e in G such that the closure of $\tilde{U}_o(H)$ in G is contained in H, then $\mathfrak{h}$ is a closed linear subspace of $\mathfrak{g}$ and a Lie subalgebra of $\mathfrak{g}$.

Remark If H is a closed subgroup, then the above condition is satisfied.

If H in the above theorem is locally compact, we can get a stronger result by using the big theorem that any locally compact group without small group is a Lie group. (Cf. [24] p175.) Namely, we can prove the following :

1.4.2 Theorem <u>A strong ILB-Lie group has no small subgroup, that is, there is a neighborhood of the identity which does not contain any non-trivial subgroup.</u>

<u>Remark</u> The above theorem is not a specific character of strong ILB-Lie groups. In fact, we can prove by a similar method that any ILB-Lie group has no small subgroup.

It is not known yet the existence of a strong ILB-Lie subgroup having a prescribed subalgebra as the Lie algebra. This can be proved only under very strong conditions. (Cf. §VII.)

The next Theorem shows the uniqueness of ILB-Lie subgroups :

1.4.3 Theorem <u>Let</u> G <u>be a strong ILB-Lie group modeled on a Sobolev chain</u> $\{\mathbb{E}, \mathbb{E}^k, k \in N(d)\}$, <u>and let</u> $\mathfrak{g}$ <u>be its Lie algebra. Suppose</u> H_1, H_2 <u>are connected ILB-Lie subgroups of</u> G <u>with the same Lie algebra</u> $\mathfrak{h}$ <u>as a subalgebra of</u> $\mathfrak{g}$. <u>Then, we have</u> $H_1 = H_2$.

<u>Proof of 1.4.1.</u>

We keep the notation and assumptions as in the statement of 1.4.1. Let $c(t)$, $t \in [0,1]$, is a C^1-curve in G with $c(0) = e$. Namely, c is a C^1-curve in G^k for every $k \in N(d)$. $\dot{c}(t)$ means the tangent vector of $c(t)$.

1.4.4 Lemma <u>If</u> $c(t) \in H$ <u>for any t, then</u> $\exp t\dot{c}(0) \in H$.

Proof. Define piecewise C^1-curves $c_n(t)$ as follows :

$$c_n(t) = \begin{cases} c(t) & : t \in [0, \frac{1}{n}) \\ c(t-\frac{m}{n})c(\frac{1}{n})^m & : t \in [\frac{m}{n}, \frac{m+1}{n}),\ m = 1,2,\dots \end{cases}$$

Obviously, $c_n(0) \in H$ for any t and satisfies $c_n(t)c_n(\frac{m}{n}) = c_n(t+\frac{m}{n})$. Let (U,ξ) be an ILB-coordinate of G at e. We put $C(t) = \xi^{-1}c(t)$, $C_n(t) = \xi^{-1}c_n(t)$. Let $\tilde{X}$ be the right invariant vector field defined by $\tilde{X} = dR_g\dot{c}(0)$. Since $\dot{c}(0) \in \mathfrak{g}$, $\tilde{X}$ is a smooth vector field on G^k for every $k \in N(d)$. We put $X = d\xi^{-1}\tilde{X}$.

For any $k \geq d+1$, there is a δ_k-neighborhood $W(\delta_k)$ of 0 in $\mathfrak{g}^k$ such that $\|(dX)_u w\|_{k'} \leq K\|w\|_{k'}$, $\|\theta(w,u,v)\|_{k'} \leq K\|w\|_{k'}$, $k' = k, k-1$, (cf.1.2.5) and $\|(d_2\theta)(w,u,v)(u')\|_{k-1} \leq K\|w\|_k\|u'\|_k$ (cf. (N,5)) for any $u, v \in W(\delta_k)$, where K is a positive constant and $(d_2\theta)(w,u,v)$ means the derivative at (w,u,v) with respect to the second variable. By the second inequality, we have $\|\eta(u,v) - \eta(u',v)\|_k \leq K\|u - u'\|_k$, hence for sufficiently large n,

$$\|C_n(t)\|_k \leq m_t K \max_{0\leq s<\frac{1}{n}} \|C(s)\|_k \leq \frac{m_t}{n} K(\|\dot{c}(0)\|_k + 1),$$

where m_t is the integer such that $0 \leq t - \frac{m_t}{n} < \frac{1}{n}$. Therefore, $C_n(t)$ is contained in $W(\delta_k)$ for sufficiently small t, say $0 \leq t < \varepsilon_k$. Remark that ε_k does not depend on n. Let D^+ denote the derivative from the right hand side. Then, using the inequality for $(d_2\theta)$, we have

$$\| D^+C_n(t) - D^+C_n(\frac{m_t}{n})\|_{k-1} \leq K'\|\dot{C}_n(t - \frac{m_t}{n}) - \dot{C}(0)\|_k ,$$

where K' is a positive constant.

Moreover, for sufficiently large n,

$$\|X(C_n(t)) - X(C_n(\frac{m_t}{n}))\|_k \leq K^2\|C(t - \frac{m_t}{n})\|_k \leq \frac{1}{n}K^2(\|\dot{c}(0)\|_k + 1), \quad 0 \leq t < \varepsilon_k .$$

Since $D^+C_n(\frac{m}{n}) = X(C_n(\frac{m}{n}))$, we see $\lim_{n\to\infty} \|D^+C_n(t) - X(C_n(t))\|_{k-1} = 0$ for $t < \varepsilon_k$. Thus, $\lim_{n\to\infty} \|C_n(t) - \int_0^t X(C_n(t'))dt'\|_{k-1} = 0$ for $0 \leq t < \varepsilon_k$.

Let $C_0(t) = \xi^{-1}\exp t\dot{c}(0)$. We have easily

$$\lim_{n\to\infty} \|C_0(t) - C_n(t) - \int_0^t (X(C_0(t')) - X(C_n(t'))dt'\|_{k-1} = 0, \quad 0 \leq t < \varepsilon_k .$$

On the other hand,

$$\|C_0(t) - C_n(t) - \int_0^t (X(C_0(t')) - X(C_n(t')))dt'\|_{k-1} \geq (1 - Kt) \|C_0(t) - C_n(t)\|_{k-1}$$

Thus $c_n(t)$ converges to $\exp t\dot{c}(0)$ in G^{k-1} for sufficiently small t. However, the convergence for small t implies the convergence for all t by virtue of the equality $c_n(t)c_n(\frac{m}{n}) = c_n(t+\frac{m}{n})$. Concequently, $c_n(t)$ converges to $\exp t\dot{c}(0)$ in G^k for any t and for any $k \in N(d)$.

Let $\tilde{U}$ be the neighborhood of e of G such that the closure of $\tilde{U}_o(H)$ in G is contained in H. For sufficiently small t, every $c_n(t)$ is contained in $\tilde{U}$ and hence in $\tilde{U}_o(H)$. Thus, $\exp t\dot{c}(0) \in H$.

1.4.1 Theorem is now an easy application of the above lemma. Let $a(t)$, $b(t)$ be one parameter subgroups in H with the initial tangent X, Y respectively. $c(t) = a(t)b(t)$ is a smooth curve in G with the initial tangent $X + Y$ and $c(t) \in H$ for all t. By the above lemma, $\exp t(X+Y) \in H$. Thus, we see that $\mathfrak{h}$ is a linear space.

Let $X_n \in \mathfrak{h}$ be a sequence converging to X in $\mathfrak{g}$. Then, for every t and for every $k \in N(d)$, $\exp tX_n$ converges to $\exp tX$ in G^k. Let $\tilde{U}$ be as above. Then, there is a positive number ε independent from n such that $\exp tX_n \in \tilde{U}$ for every $t \in [0,\varepsilon)$, hence $\exp tX_n \in \tilde{U}_o(H)$. Thus, by the assumption for $\tilde{U}$, we have $\exp tX \in \tilde{U}_o(H) \subset H$. Therefore, $\mathfrak{h}$ is a closed subspace.

For any $g \in H$, we see that $Ad(g)\mathfrak{h} = \mathfrak{h}$, where $Ad(g) = \frac{d}{dt}\Big|_o g \exp tu\, g^{-1}$. The Lie bracket $[u,v]$ is given by $\frac{d}{dt}\Big|_o Ad(\exp tv)u$. (It is easy to see that this Lie bracket is equal to $(\widetilde{\tilde{u}\tilde{v} - \tilde{v}\tilde{u}})(e)$.) Therefore, $\mathfrak{h}$ is a Lie subalgebra.

Proof of 1.4.2.

We keep the notations as in 1.2.1. By the local property (N,3), we see that

$$\eta \; : \; V \cap E^{k+1} \times V \cap E^k \longmapsto U \cap E^k$$

is a C^1-mapping. Thus, there are open neighborhoods W^{k+1}, W^k of 0 of $V \cap E^{k+1}$, $V \cap E^k$ respectively such that $\|(d\rho_u)_v w - w\|_k \leq \varepsilon \|w\|_k$, $\varepsilon < 1$, for every $u \in W^{k+1}$, $v \in W^k$, where $\rho_u(v) = \eta(u,v)$. We have only to show the following :

1.4.5 Lemma W^{k+1} has no non-trivial subgroup.

Proof. Put $\eta_v(u) = \eta(u,v)$. Then, we have

$$\eta_u^m(u) - \eta_u^{m-1}(u) = \int_o^1 (d\rho_{\eta_u^{m-1}(u)})_{tu}\, u\, dt,$$

hence,

$$\eta_u^m(u) = u + \sum_{k=o}^{m-1} \int_o^1 (d\rho_{\eta_u^k(u)})_{tu}\, u\, dt = mu + \sum_{k=o}^{m-1} \int_o^1 \{ (d\rho_{\eta_u^k(u)})_{tu} - I \}\, u\, dt.$$

Thus, $\|\eta_u^m(u)\|_k \geq m(1-\varepsilon)\|u\|_k$. This implies that if $\eta_u^m(u)$ is contained in W^{k+1} for all m, then u must be 0.

Proof of 1.4.3

It is enough to show that $H_1 \subset H_2$, because we have $H_2 \subset H_1$ by interchanging H_1 and H_2.

Let h_t be a smooth curve in H_1 starting at the identity e. Let $u_t = dR_{h_t^{-1}} \frac{d}{dt} h_t$. Then, u_t is contained in the Lie algebra $\mathfrak{h}_1$ of H_1 and $t \mapsto u_t$ is a smooth mapping of $[0,1]$ into $\mathfrak{h}_1$. Let $\tilde{u}_t$ be the (time dependent) smooth vector field defined by $\tilde{u}_t(g) = dR_g u_t$. This is a smooth vector field on G^k for every $k \in N(d)$. Since $\mathfrak{h}_1 = \mathfrak{h}_2$ and H_2 is a strong ILB-Lie subgroup of G, we see that the restriction of $\tilde{u}_t$ on H_2^k is also a smooth vector field on H_2^k. Thus, the integral curve h_t' of $\tilde{u}_t$ through the identity is contained in H_2^k for every k. On the other hand, the uniqueness of the integral curve of $\tilde{u}_t$ implies that $h_t = h'_t$. Thus we have $H_1 \subset H_2$.

I.5 Invariant connections on strong ILB-Lie groups.

Remark at the first stage that to give a connection on a finite dimensional manifold is to give a normal coordinate at every point. Namely, suppose $\tilde{W}$ be a neighborhood of the 0-section of the tangent bundle T_X of X. We denote by $\tilde{W}_x$ the intersection of $\tilde{W}$ and the tangent space T_xX at $x \in X$. Suppose furthermore that there is a C^∞-diffeomorphism ξ of $\tilde{W}$ onto an open neighborhood of the diagonal of $X \times X$, such that $\tilde{\xi}$(0-section) = diagonal and $p_1\tilde{\xi}y = \pi y$, where $p_1(x,y) = x$ and $\pi : T_X \mapsto X$ is the projection.

Then, by regarding $(\tilde{W}_x, \tilde{\xi}|\pi^{-1}(x))$ as a normal coordinate at x, we can define a C^∞-connection on X. Namely, the covariant derivative at x is just the same as the ordinary differential at x with respect to the normal coordinate.

Keep this fact in mind. We use the same method in the definition of an invariant connection on a strong ILB-Lie group.

Let G be a strong ILB-Lie group modeled on $\{\mathfrak{g}, \mathfrak{g}^k, k \in N(d)\}$. $\mathfrak{g}^k$ is naturally identified with the tangent space of G^k at e and $\mathfrak{g} = \cap \mathfrak{g}^k$ is identified with the tangent space of G at e, (or simply, we call $\mathfrak{g}$ the tangent space of G at e.)

Let T_{G^k} be the tangent bundle of G^k and T_G the inverse limit of T_{G^k}, i.e. $T_G = \cap T_{G^k}$ with the inverse limit topology. We call T_G the tangent bundle of G. (In fact, T_G is a vector bundle in a sense of Frechet manifold.) Let $\pi^k : T_{G^k} \mapsto G^k$ be the projection. Then, it is clear that π^k can be extended to the projection $T_{G^{k-1}} \mapsto G^{k-1}$. So, in this sense we can omit the suffix k of π. Taking the inverse limit, we also have the projection $\pi : T_G \mapsto G$. (So, we can say this π can be extended to the projection of T_{G^k} onto G^k.)

Let (U,ξ) be an ILB-coordinate of G at e. (Remark that U is an open neighborhood of 0 of $\mathfrak{g}^d$.) Using (G,7), we see that $\tilde{U} = dR(U \times G^d)$ is an open neighborhood of 0-section of T_{G^d} and $\tilde{U} \cap T_{G^k}$ (resp. $\tilde{U} \cap T_G$) is an open neighborhood of 0-section of T_{G^k} (resp. T_G). We denote by $\tilde{U}_g$ the intersection of $\tilde{U}$ and the tangent space T_gG^d of G^d at g. Then, obviously $\tilde{U}_g \cap T_{G^k} = \tilde{U} \cap T_gG^k$, if $g \in G^k$.

Let $\hat{U} = \cup \{(g, R_g\xi(U)) : g \in G^d\}$. Then $\hat{U} \cap (G^k \times G^k)$ is an open right-invariant neighborhood of the diagonal of $G^k \times G^k$ for any $k \in N(d)$, hence $\hat{U} \cap (G \times G)$ is an open neighborhood of the diagonal of $G \times G$ which is invariant by right translations.

Define a mapping $\tilde{\xi} : \tilde{U} \cap T_G \mapsto \hat{U} \cap (G \times G)$ by $\tilde{\xi}(u) = (\pi u, \xi(dR_g^{-1}u)\cdot g)$, $\pi u = g$, and $\tilde{\xi}_g : \tilde{U} \cap T_gG \mapsto R_g\xi(U \cap \mathfrak{g}) = R_g\xi(U) \cap G$ by $\tilde{\xi}_g(u) = \xi(dR_g^{-1}u)\cdot g$. Then, we see easily that $\tilde{\xi}_{gh}(dR_h u) = R_h\tilde{\xi}_g(u)$, and using (G,7) again, we see that $\tilde{\xi}$ can be extended to a homeomorphism of $\tilde{U} \cap T_{G^k}$ onto $\hat{U} \cap (G^k \times G^k)$. Let $p_1 : G \times G \mapsto G$ be the projection such that $p_1(x,y) = x$. Obviously, p_1 can be extended to the projection of $G^k \times G^k$ onto G^k. By definition, we have $p_1\tilde{\xi}(u) = \pi u$.

Now, we assume the following :

(G,9) $\tilde{\xi} : \tilde{U} \cap T_G \mapsto \hat{U} \cap (G \times G)$ can be extended to a C^∞-diffeomorphism of $\tilde{U} \cap T_{G^k}$ onto $\hat{U} \cap (G^k \times G^k)$.

Thus, the following is now easy to prove :

1.5.1 Proposition _If G satisfies the condition (G,9), then there exists a smooth connection ∇^k on each G^k such that ∇^k is invariant under right translations_ (i.e. $\nabla_{dR_h u} dR_h \tilde{v} = dR_h \nabla_u \tilde{v}$.) _and ∇^{k-1} is the extension of ∇^k._ (_In this sense, we can omit the suffix k of ∇._) _Taking the inverse limit of ∇, we call this a connection on G, which, we may say, can be extended to a smooth connection on every G^k._

Let ∇ be a connection on G given by the above manner. Let $c(t)$ be a curve in G such that $c : [0,1] \mapsto G^k$ is a smooth mapping for every $k \in N(d)$. For any $u \in T_{c(0)}G$, since this is regarded as an element of $T_{c(0)}G^k$ for every $k \in N(d)$, we can define the parallel displacement $u(t)$ of u along the curve $c(t)$. However, $u(t)$ does not depend on k. Thus, $u(t)$ is in fact contained in T_G. We call $u(t)$ the parallel displacement of u along the curve $c(t)$.

Now, a curve $c(t)$ in G will be called a geodesic, if $c(t)$ is geodesic in every G^k. However, it is not clear that there exist a geodesic in G, because of the following reason :

For any $u \in \mathfrak{g}$ and $k \in N(d)$, there is a geodesic $c(t)$ in G^k defined on $[0,t_k)$. However, the number t_k may depend on k in general. So, if $t_k \mapsto 0$ with $k \mapsto \infty$, then we can not conclude the existence of the geodesic in G with the initial direction u. If $t_d = t_k$ for every $k \in N(d)$, we say that the geodesic $c(t)$ has a _regularity._ To ensure such property, we have to assume a stronger condition than

(G,9) above. (Cf. I.6)

Moreover, (G,9) is not independent from (G,3) - (G,7). We discuss this at first.

1.5.2 Lemma <u>The local expression of</u> $\tilde{\xi}$ <u>is given by</u> $(u, \eta(\theta(u,v,\iota(v)), v))$.

Proof. Let W be a neighborhood of 0 in $\mathfrak{g}^d$ such that $W \subset U$. Let $\xi_g : W \cap \mathfrak{g} \mapsto \xi(W \cap \mathfrak{g})\cdot g$ be the mapping defined by $\xi_g(u) = \xi(u)\cdot g$. This mapping is regarded as a smooth chart of G around g. Denote by $T_{\xi(W\cap\mathfrak{g}^k)\cdot g}$ the tangent bundle of $\xi(W \cap \mathfrak{g}^k)\cdot g$. Then, the mapping $d\xi_g : W \cap \mathfrak{g}^k \times \mathfrak{g}^k \mapsto T_{\xi(W\cap\mathfrak{g}^k)\cdot g}$ can be regarded as a local coordintate (and also a local trivialization) of T_{G^k} for every $k \in N(d)$. Let Ω be the inverse image of $\hat{U} \cap T_{\xi(W\cap\mathfrak{g}^d)\cdot g}$ by $d\xi_g$. This is an open neighborhood of $(0,0)$ in $W \cap \mathfrak{g}^d \times \mathfrak{g}^d$.

The mapping $(\xi_g, \xi_g) : W\cap\mathfrak{g} \times W\cap\mathfrak{g} \mapsto \xi(W\cap\mathfrak{g})\cdot g \times \xi(W\cap\mathfrak{g})\cdot g$ is an ILB-local coordinate of $G \times G$ at (g,g).

Thus, to obtain the local expression of $\tilde{\xi}$, we have to investigate

$$(\xi_g, \xi_g)^{-1}\, \tilde{\xi}\, d\xi_g \ .$$

Let $(u,v) \in \Omega \cap (\mathfrak{g} \times \mathfrak{g})$. Then, $(\xi_g,\xi_g)^{-1}\tilde{\xi}(d\xi_g)_u v = (u, \xi_g^{-1}\tilde{\xi}_{\xi(u)g}(d\xi_g)_u v)$,

and $\xi_g^{-1}\tilde{\xi}_{\xi(u)g}(d\xi_g)_u v = \xi_g^{-1}\xi_{\xi(u)g}\, dR_{\xi(u)^{-1}}\,(d\xi)_u v = \eta(dR_{\xi(u)^{-1}}(d\xi)_u v, u)$.

On the other hand,

$$\theta(w,u,v') = (d\eta_{v'})_u w = (d\xi)^{-1} dR_{\xi(v')}(d\xi)_u w \ .$$

So, put $v' = \iota(u)$, (i.e. $\xi(v') = \xi(u)^{-1}$.) Then, $dR_{\xi(u)^{-1}}(d\xi)_u w$ is an element of $\mathfrak{g}$. Since $(d\xi)_e : \mathfrak{g} \mapsto \mathfrak{g}$ is the identity, we have that

$$\theta(w,u,\iota(u)) = dR_{\xi(u)^{-1}}(d\xi)_u w.$$

Thus, we have the desired result.

Let $\Xi(v,u) = \eta(\theta(v,u,\iota(u)),u)$. Then the condition (G,9) imlpies the following

(N,8) $\Xi : \Omega \cap (\mathfrak{g} \times \mathfrak{g}) \mapsto \mathfrak{g}$ can be extended to a C^∞-mapping of $\Omega \cap (\mathfrak{g}^k \times \mathfrak{g}^k)$ into

$\mathfrak{g}^k$ for every $k \in N(d)$.

1.5.3 Lemma $\eta(u,v) = \Xi(\theta(u,0,v),v)$ and $\theta(w,u,v) = (d_1\Xi)_{(\theta(u,0,v),v)}\theta(w,0,v)$, where $d_1\Xi$ means the derivative with respect to the first variable.

Proof. By definition, we see $\Xi(\theta(u,0,v),v) = \eta(\theta(\theta(u,0,v),v,\iota(v)),v)$. On the other hand, by the associative law, we have

$$\eta(\eta(u,v),w) = \eta(u, \eta(v,w)).$$

Taking derivatives,

$$\theta(\theta(w,u,v),\eta(u,v),v') = \theta(w, u, \eta(v,v')).$$

Especially, $\theta(\theta(u,0,v),v,\iota(v)) = \theta(u,0,0) = u$. Thus, we see the first equality.

Taking derivatives of the first equality, we have the second one.

By this Lemma, we see easily the following :

1.5.4 Proposition The conditions (N,9) and (N,ζ) stated below imply (N,3), (N,4) and (N,5).

(N,ζ) Letting $\zeta(u,v) = (d\eta_v)_0 u$, $\zeta : \mathfrak{g} \times W \cap \mathfrak{g} \mapsto \mathfrak{g}$ can be extended to a C^ℓ-mapping of $\mathfrak{g}^{k+\ell} \times W \cap \mathfrak{g}^k$ into $\mathfrak{g}^k$ for any $k \in N(d)$ and $\ell \geq 0$.

I.6 Regularities of connections.

1.6.1 Lemma A local expression of the above connection ∇ on G is given at the identity by $\Gamma(w)(u,v) = -(d_1^2\Xi)_{(0,w)}(u,v)$.

Proof. Notations are as above, since ∇ is invariant under right translations, we have only to consider a local expression of ∇ on a neighborhood of the identity. A local expression of a connection is given by

$$\Gamma(w)(u,v) = d\xi^{-1}\nabla_{(d\xi)_w u}\, d\xi\, v\ .$$

This is the same content of a local expression Γ^i_{jk} of a connection on a finite dimensional manifold.

Now, recall that ∇ is defined by the ordinary differentiation using a prescribed normal coordinate.

$$\Gamma(w)(u,v) = d\xi^{-1}\lim_{s\to 0}\frac{1}{s}\{(d\tilde{\xi}_{\xi(w)}^{-1}d\xi\, v)((d\xi)_w\, su) - (d\tilde{\xi}_{\xi(w)}^{-1}d\xi\, v)(0)\}$$

$$= \lim_{s\to 0}\frac{1}{s}\{(d\xi^{-1}d\tilde{\xi}_{\xi(w)}^{-1}d\xi\, v)(w+su) - (d\xi^{-1}d\tilde{\xi}_{\xi(w)}^{-1}d\xi\, v)(w)\}.$$

On the other hand, we see

$$(d\tilde{\xi}_{\xi(w)})_u v = dR_{\xi(w)}(d\xi)_\gamma\, dR_{\xi(w)}^{-1}v, \quad \gamma = dR_{\xi(w)}^{-1}u.$$

Thus, $d\xi^{-1}d\tilde{\xi}_{\xi(w)}^{-1}d\xi\, v = d\xi^{-1}dR_{\xi(w)}d\xi^{-1}dR_{\xi(w)}^{-1}d\xi\, v$. Since $\theta(w,u,v) = d\xi^{-1}dR_{\xi(v)}(d\xi)_u v$, we have $d\xi^{-1}d\tilde{\xi}_{\xi(w)}^{-1}(d\xi)_{w'}v = \tilde{\theta}(0,w)\tilde{\theta}(w',\iota(w))v$, where $\tilde{\theta}(*,!)w = \theta(w,*,!)$. Hence, this is equal to $\tilde{\theta}(0,w)(d_1\Xi)_{(\tilde{\theta}(0,\iota(w)w',\iota(w))}\tilde{\theta}(0,\iota(w))v$. (Cf. 1.5.3.)

Thus, we have

$$\Gamma(w)(u,v) = \tilde{\theta}(0,w)(d_1^2\Xi)_{(\tilde{\theta}(0,\iota(w))w,\iota(w))}(\tilde{\theta}(0,\iota(w)u,\tilde{\theta}(0,\iota(w))v).$$

On the other hand, since $\eta(\eta(u,v),w) = \eta(u,\eta(v,w))$, taking derivatives, we see

$$\tilde{\theta}(\eta(u,v),v')\tilde{\theta}(u,v) = \tilde{\theta}(u,\eta(v,v')).$$

Insert the second formula of 1.5.3 into the above equality. Then,

$$(d_1\Xi)_{(\tilde{\theta}(0,v')\eta(u,v),v')}\tilde{\theta}(0,v')(d_1\Xi)_{(\tilde{\theta}(0,v)u,v)}\tilde{\theta}(0,v)w$$

$$= (d_1\Xi)_{(\tilde{\theta}(0,\eta(v,v'))u,\eta(v,v'))}\tilde{\theta}(0,\eta(v,v'))w.$$

Change the variable $v'\mapsto w$, $v\mapsto\iota(w)$, $u\mapsto w'$, $w\mapsto v$. Then, we have

$$(d_1\Xi)_{(\tilde{\theta}(0,w)\eta(w',\iota(w)),w)}\tilde{\theta}(0,w)(d_1\Xi)_{(\tilde{\theta}(0,\iota(w))w',\iota(w))}\tilde{\theta}(0,\iota(w))v = v.$$

Take the derivative with respect to the variable w' at w. Thus, we have

$$(d_1^2\Xi)_{(0,w)}(\tilde{\theta}(0,w)(d\eta_{\iota(w)})_w u,\ \tilde{\theta}(0,w)(d_1\Xi)_{(\tilde{\theta}(0,\iota(w))w,\ \iota(w))}\tilde{\theta}(0,\iota(w))v)$$

$$+ (d_1\Xi)_{(0,w)}\Gamma(w)(u,v) = 0.$$

Remark that $(d_1\Xi)_{(0,w)} = \mathrm{id.}$ and $\tilde{\theta}(w,0) = \mathrm{id.}$ and that

$$\tilde{\theta}(0,w)(d_1\Xi)_{(\tilde{\theta}(0,\iota(w))w,\ \iota(w))}\tilde{\theta}(0,\iota(w))v = \tilde{\theta}(0,w)\tilde{\theta}(w,\ \iota(w))v = v.$$

Then, we have the required equality.

Now, we discuss about the following question : When the exponential mapping, Exp, is defined and gives an ILB-coordinate of G ?

This question is not completely solved. First of all, we have to restrict our groups in strongly LIH-Lie groups. (We have to use the differentiability of norms.)

Now, let $\{ \mathbb{E}, E^k, k \in N(d)\}$ be a Sobolev chain such that every E^k is a Hilbert spase. Let U be an open neighborhood of 0 of E^d.

A connection ∇ on $U \cap \mathbb{E}$ (in a sense of Frechet manifold) is called an ILH-connection, if ∇ can be extended to a smooth connection ∇ on every $U \cap E^k$. Since $U \cap \mathbb{E}$ is an open subset of linear space, ∇ defines an $\mathbb{E}$-valued bi-linear form Γ by $\Gamma(x)(u,v) = \nabla_u v$ at x, where u, v in the right hand side are identified with vector fields obtained by the parallel translation using the linear structure of $\mathbb{E}$. Since ∇ is an ILH-connection, the bi-linear form Γ can be extended to a smooth E^k-valued bilinear form of $U \cap E^k$. As a matter of course, the equation of a geodesic is given by

$$x''(t) + \Gamma(x(t))(x'(t), x'(t)) = 0.$$

Therefore, if $x(t)$ is geodesic in $U \cap E^k$, then so it is in $U \cap E^{k-1}$.

Now assume the following inequalities (I) and (II) :

Let $\| \ \|_k$ be the norm of E^k, that is, $\|u\|_k^2 = < u, u >_k$ using the inner product of E^k.

(I) $$\|\Gamma(x)(u,v)\|_k \leq C\{\|u\|_k\|v\|_d + \|u\|_d\|v\|_k\} + C^2\|x\|_k\|u\|_d\|v\|_d + P_k(\|x\|_{k-1})\|u\|_{k-1}\cdot\|v\|_{k-1}, \quad k > d,$$

where C is a constant which does not depend on k and P_k is a polynomial with positive coefficients.

Let $d\Gamma(x)(w,u,v)$ be the derivative of Γ at x with respect to the first variable x.

(II) $$\|d\Gamma(x)(w,u,v)\|_k \leq C^2\{\|u\|_k\|v\|_d\|w\|_d + \|u\|_d\|v\|_k\|w\|_d + \|u\|_d\|v\|_d\|w\|_k\} + P_k'(\|x\|_k)\|u\|_{k-1}\|v\|_{k-1}\|w\|_{k-1}, \quad k > d,$$

where C is the same constant as in (I) and P'_k is a positive continuous function of $\|x\|_k$. P_k and P'_k may depend on k. The constant C^2 used in (I) and (II) are only for convenience of notations. They can be replaced by some other constants as far as they do not depend on k.

Now, let Exp be the exponential mapping at the origin. Since ∇ is an ILH-connection, there is an open neighborhood W^k for each k such that Exp is a smooth diffeomorphism of W^k onto an open neighborhood $\mathrm{Exp}\,W^k$ of 0 in E^k. In general, there is no relation between W^k and W^{k+1}. However, by virtue of inequalities (I) and (II), we have the following theorem. This was the main theorem of [29].

1.6.2 Theorem <u>Suppose an ILH-connection ∇ on $U \cap \mathbb{E}$ satisfies the inequalities (I) and (II). Let V be an open neighborhood of 0 in E^d on which the exponential mapping can be defined. Then, this mapping can be defined also on $V \cap E^k$ as the exponential mapping with respect to the extended connection ∇ on $U \cap E^k$.</u>

<u>Moreover, there exists an open neighborhood $\tilde{V}$ of 0 in E^d such that</u>

(a) $\mathrm{Exp}(\tilde{V} \cap E^k) = (\mathrm{Exp}\tilde{V}) \cap E^k$,

(b) <u>letting $\tilde{V}^k = (\mathrm{Exp}\tilde{V}) \cap E^k$, the exponential mapping is a smooth diffeomorphism of $\tilde{V} \cap E^k$ onto $\tilde{V}^k$.</u>

The above theorem was proved in [29] by the following manner : It is easy to see that there exists an open neighborhood V on 0 in E^d on which Exp is defined. (Cf. [8].)

At first, we have to prove the regularity of geodesics. For the proof of this, we need only the inequality (I) and moreover the constant C may depend on k. The regularity of geodesics implies $\mathrm{Exp}(V \cap E^k) \subset (\mathrm{Exp}V) \cap E^k$. Obviously, there exists an open neighborhood $\tilde{V}$ of 0 in E^d such that $\mathrm{Exp} : \tilde{V} \longmapsto \mathrm{Exp}\tilde{V}$ is diffeomorphic. We may assume that $\tilde{V}$ is a star shaped neighborhood of 0.

Secondly, we have to prove the openness of the exponential mapping

$$\mathrm{Exp} : \tilde{V} \cap E^k \longmapsto (\mathrm{Exp}\,\tilde{V}) \cap E^k.$$

This was done by proving the derivative dExp at u is an isomorphism for every $u \in$

$\tilde{V} \cap E^k$. This implies that Exp is a diffeomorphism of $\tilde{V} \cap E^k$ onto an open neighborhood of 0 in $(\mathrm{Exp}\,\tilde{V}) \cap E^k$. However this does not imply $\mathrm{Exp}(\tilde{V} \cap E^k) = (\mathrm{Exp}\,\tilde{V}) \cap E^k$ in general.

So, thirdly, we have to prove the closedness of the image $\mathrm{Exp}(\tilde{V} \cap E^k)$ in $(\mathrm{Exp}\,\tilde{V}) \cap E^k$. Since the connectedness of $\tilde{V}$ implies $(\mathrm{Exp}\,\tilde{V}) \cap E^k$ is connected for any k, these three parts complete the proof.

To prove these three parts, we had to establish some comparison theorems of the solutions of the geodesic equation. We used differential inequalities with respect to $\|x(t)\|_k^2$. So, this method can not be applied for ILB-connections. For details, see the previous paper [29]. The basic idea of the proof of the inequalities (I) and (II) will be given in the next chapter.

§ II Groups of diffeomorphisms

II.1. Some remarks on local properties of groups of diffeomorphisms.

The goal here is the proof that the group of diffeomorphisms is a strong ILB- and a strong ILH- Lie group.

Let M be a smooth and closed manifold and T_M its tangent bundle. By $\Gamma(T_M)$ and $\gamma^k(T_M)$, we denote the space of all C^∞-vector fields with C^∞-topology and all C^k-vector fields with C^k-topology respectively. Correspondingly, by $\mathcal{D}$ and $\hat{\mathcal{D}}^k$ we denote the group of all C^∞-diffeomorphisms with C^∞-topology and the group of all C^k-diffeomorphisms with C^k-topology. (As M is fixed in this section, we do not use the notations $\mathcal{D}(M)$ and $\hat{\mathcal{D}}(M)$.)

Consider a smooth connection on M and let Exp be its exponential mapping. Since M is compact, the following is easy to prove :

2.1.1 Lemma <u>There exists an open neighborhood $\mathfrak{U}$ of 0 in $\gamma^1(T_M)$ such that the mapping $\xi(u)$ defined by $\xi(u)(x) = \operatorname{Exp} u(x)$ for $u \in \mathfrak{U}$ is a C^1-diffeomorphism of M, and $\xi : \mathfrak{U} \longmapsto \hat{\mathcal{D}}^1$ is an into-homeomorphism. Moreover, if $u \in \mathfrak{U} \cap \Gamma(T_M)$, then $\xi(u) \in \mathcal{D}$, and $\xi : \mathfrak{U} \cap \Gamma(T_M) \longmapsto \mathcal{D}$ is an into-homeomorphism.</u>

We use the above $(\mathfrak{U},\xi)$ as an ILB- or ILH- coordinate of $\mathcal{D}$ at the identity. Obviously, we may change the connection, getting another ILB- or ILH- coordinate. However, this makes no difference on strong ILB- or stron ILH- Lie group structures on $\mathcal{D}$. This will be discussed in § II.

Since $\hat{\mathcal{D}}^1$ is a topological group, there is an open neighborhood $\mathfrak{V}$ of 0 in $\gamma^1(T_M)$ such that $\xi(\mathfrak{V}) = \xi(\mathfrak{V})^{-1}$ and $\xi(\mathfrak{V})^2 \subset \xi(\mathfrak{U})$, hence $\xi(\mathfrak{V} \cap \Gamma(T_M)) = \xi(\mathfrak{V} \cap \Gamma(T_M))^{-1}$ and $\xi(\mathfrak{V} \cap \Gamma(T_M))^2 \subset \xi(\mathfrak{U} \cap \Gamma(T_M))$.

Let $\eta(u,v) = \xi^{-1}(\xi(u)\xi(v))$. Then, we have

$$\eta(u,v)(x) = \operatorname{Exp}_x^{-1}\operatorname{Exp} u(\operatorname{Exp}_x v(x)) \qquad \text{..............} \ (1).$$

Moreover, if $d\eta_v$ is defined in any sense, this should be given by

$$\theta(w,u,v)(x) = (d\eta_v)_u w\ (x) = (dExp_x)^{-1}(dExp)_{u(Exp_x v(x))} w(Exp_x v(x)) \quad \ldots\ldots\ (2).$$

Hence, using $Exp\ \iota(v)(Exp\ v(x)) = x$,

$$\Xi(u,v)(x) = Exp_x^{-1} Exp(dExp_x)_{v(x)} u(x) \quad \ldots\ldots\ (3).$$

Let $\zeta(u,v) = (d\eta_v)_o u$. Then,

$$\zeta(u,v)(x) = (dExp_x)^{-1} u(Exp_x v(x)) \quad \ldots\ldots\ (4),$$

and hence

$$\eta(u,v) = \Xi(\zeta(u,v),v) \quad \ldots\ldots\ (5),$$

$$\theta(w,u,v) = (d_1\Xi)_{(\zeta(u,v),v)} \zeta(w,v) \quad \ldots\ldots\ (6).$$

So to consider the group $\mathcal{D}$, we have mainly to concern about the mappings Ξ and ζ, but these are already discussed in [30]. We repeat here three fundamental lemmas, with short comments but without proofs.

Let E be a smooth finite dimensional riemannian vector bundle over M. $\Gamma(E)$ mean the space of all smooth sections of E with the C^∞-topology. Let $\tau(Exp_x u(x))$ be the parallel displacement along the curve $Exp_x tu(x)$, $t \in [0,1]$. For any $u \in \hat{U} \cap \Gamma(T_M)$ (cf. 2.1.1.) and $v \in \Gamma(E)$, we put $R'(v,u) = \tau(Exp_x u(x))^{-1} v(Exp_x u(x))$. This is a mapping of $\Gamma(E) \times \hat{U} \cap \Gamma(T_M)$ into $\Gamma(E)$.

We define two norms (B-norm and H-norm) on $\Gamma(E)$ as follows :

(B-norm): $\|u\|_k = \max\{|\nabla^s u|(x) : x \in M,\ s \leq k\}$,

(H-norm): $\|u\|_k = < u,u >^{\frac{1}{2}}$, where

$$< u,v > = \sum_{s=0}^{k} \int_M < \nabla^s u, \nabla^s v > dV.$$

Let $\gamma^k(E)$ be the completion of $\Gamma(E)$ with respect to the B-norm $\| \ \|_k$. Namely, $\gamma^k(E)$ is the space of all C^k-sections with C^k-uniform norm.

Let $\Gamma^k(E)$ be the completion of $\Gamma(E)$ with respect to the H-norm $\| \ \|_k$.

2.1.2 Lemma $R' : \Gamma(E) \times \hat{U} \cap \Gamma(T_M) \mapsto \Gamma(E)$ can be extended to a C^ℓ-mapping of $\Gamma^{k+\ell}(E) \times \hat{U} \cap \Gamma^k(T_M)$ into $\Gamma^k(E)$ for every $k \geq \dim M + 5$, and to a C^ℓ-mapping of $\gamma^{k+\ell}(E) \times \hat{U} \cap \gamma^k(T_M)$ into $\gamma^k(E)$ for every $k \geq 1$.

The first one was proved in Lemma 5 of [30]. The second one is much easier to prove by using Lemma 8 of [30].

Note. If E is a trivial bundle, then we may put $R'(v,u)(x) = v(\mathrm{Exp}_x u(x))$. Thus, $R'(v,u)$ is the composition of mappings, i.e. $R'(v,u) = v\xi(u)$. The notation R' means a local expression of the right translation.

Let W be a relatively compact tubular neighborhood of zero section of E, and let φ be a smooth fibre preserving mapping of $\bar{W}$ (the closure of W) into another finite dimensional smooth riemannian vector bundle F over M, i.e. if $Y \in E_x$ (the fibre of E at x), then $\varphi(Y) \in F_x$. Let $\Gamma(W)$ denote the subset of elements $u \in \Gamma(E)$ such that $u(x) \in W$ for every $x \in M$ and let $\Gamma^k(W)$ (resp. $\gamma^k(W)$) be the set $\{ v \in \Gamma^k(E)$ (resp. $\gamma^k(E)$) : $v(x) \in W$ for every $x \in M \}$. $\Gamma^k(W)$ is well-defined for $k \geq [\frac{1}{2}\dim M] + 1$ by virtue of Sobolev embedding theorem.

Since φ is smooth and fibre preserving, one can define a mapping $\Phi : \Gamma(W) \mapsto \Gamma(F)$ by $\Phi(u)(x) = \varphi(u(x))$. We call that the mapping Φ is defined from φ.

2.1.3 Lemma $\Phi : \Gamma(W) \mapsto \Gamma(F)$ can be extended to a smooth mapping of $\Gamma^k(W)$ (resp. $\gamma^k(W)$) into $\Gamma^k(F)$ (resp. $\gamma^k(F)$) for every $k \geq \dim M + 5$ (resp. $k \geq 1$). Moreover, the following inequality holds for $k \geq \dim M + 5$ (resp. $k \geq 1$) :

$$\|\Phi(u)\|_k \leq P(\|u\|_{k_0})\|u\|_k + Q_k(\|u\|_{k-1}), \quad k_0 = 2([\tfrac{1}{2}\dim M] + 1) \text{ (resp. } = 1,)$$

where P and Q_k are polinomials with positive coefficients and P does not depend on k.

The proof for Γ^k is a special case of Theorem A and B in [30]. The proof for γ^k is however much easier than the previous one.

Let $\iota : \hat{U} \cap \Gamma(T_M) \mapsto \hat{U} \cap \Gamma(T_M)$ be the mapping defined by $\iota(u) = \xi^{-1}(\xi(u)^{-1})$.

2.1.4 Lemma ι <u>can be extended to a continuous mapping of</u> $\hat{\mathcal{V}} \cap \Gamma^k(T_M)$ (resp. $\hat{\mathcal{V}} \cap \gamma^k(T_M)$) <u>into itself for every</u> $k \geq \dim M + 5$ (resp. $k \geq 1$).

The case of $\Gamma^k(T_M)$ is proved in 4° of [30] and the case $\gamma^k(T_M)$ is trivial, because $\mathcal{D}^k$ is a topological group.

Now, what we have to prove is the following :

2.1.5 Theorem $\mathcal{D}$ <u>is a strong ILH-</u> (resp. <u>a strong ILB-) Lie group modeled on</u> $\{ \Gamma(T_M), \Gamma^k(T_M), k \in N(\dim M + 5)\}$ (resp. $\{ \Gamma(T_M), \gamma^k(T_M), k \in N(1)\}$).

Proof. We have only to check the conditions (N,1) - (N,7). Notations are as above. (N,1) and (N,2) are trivial. (N,3) is given by the equality (5) and 2.1.2 - 3, because Ξ is defined from a smooth fiber preserving mapping. (Cf. the formula (3) or the proof of Lemma 2 of [30].) (N,4) is trivial, because of (5) and the fact that $\zeta(u,v)$ is linear, hence C^∞ with respect to the variable u. (N,5) is given by the formula (6) and 2.1.2 - 3.

If $\ell = 0$, then (N,6) is the same as 2.1.4. If $\ell \geq 1$, then (N,6) is not independent from others. In fact, (N,6) with $\ell \geq 1$ can be proved by (N,3) using the implicit function theorem on Banach manifolds.

Now, we want to prove (N,7) by using 2.1.3. Here, we will give the proof only for the case $\Gamma^k(T_M)$, but for $\gamma^k(T_M)$, we may simply change $\Gamma^k(T_M)$ by $\gamma^k(T_M)$ and $\dim M + 5$ by 1 in the proof below.

Let g be a smooth diffeomorphism of M and let $g^{-1}T_M$ be the pull back of the tangent bundle T_M of M. We have the canonical mapping $g_* : g^{-1}T_M \mapsto T_M$ such that if X is an element in the fibre at x of $g^{-1}T_M$, then g_*X is an element of the fibre of T_M at $g(x)$. We see that there is an open relatively compact tubular neighborhood W_1 of zero section of $g^{-1}T_M$ such that the mapping $a_g : \overline{W}_1 \longmapsto T_M$, $a_g(X) = \mathrm{Exp}_x^{-1} g^{-1}(\mathrm{Exp}\, g_* X)$, is well-defined. The mapping a_g is smooth and fibre preserving. Let $\tilde{A}_g : \Gamma(W_1) \longmapsto \Gamma(T_M)$ be the mapping defined from a_g, i.e. $\tilde{A}_g(u)(x) = a_g(u(x))$. Thus, 2.1.3 Lemma yields that $\tilde{A}_g$ can be extended to a smooth mapping of

$\Gamma^k(W_1)$ into $\Gamma^k(T_M)$ for every $k \geq \dim M + 5$.

On the other hand, we have a mapping $g^* : \Gamma(T_M) \mapsto \Gamma(g^{-1}T_M)$, $g^*u(x) = ug(x)$. This is a linear mapping and as g is smooth, g^* can be extended to a bounded linear operator of $\Gamma^k(T_M)$ into $\Gamma^k(g^{-1}T_M)$ for every $k \geq 0$. This is proved by a direct computation.

The mapping $A_g(u) = \xi^{-1}(g^{-1}\xi(u)g)$ is therefore written as the composition $A_g = \tilde{A}_g \cdot g^*$. Since every bounded linear operator is smooth, we see that $A_g : \Gamma(g_*W_1) \mapsto \Gamma(T_M)$ can be extended to a smooth mapping of $\Gamma^k(g_*W_1)$ into $\Gamma^k(T_M)$. As $\Gamma^k(g_*W_1)$ is an open neighborhood of 0 of $\Gamma^k(T_M)$ for every $k \geq [\frac{1}{2} \dim M] + 1$, there is an open neighborhood W' of 0 in $\Gamma^d(T_M)$, $d = \dim M + 5$, such that $W' \cap \Gamma(T_M) \subset \Gamma(g_*W_1)$. Since $A_g : W' \mapsto \Gamma^d(T_M)$ is smooth and $A_g(0) = 0$, there is an open neighborhood W of 0 of $\Gamma^d(T_M)$ such that $W \subset W'$ and $A_g(W) \subset U$, hence $A_g(W \cap \Gamma(T_M)) \subset U \cap \Gamma(T_M)$. This complete the proof.

Remark. Since the group $\mathcal{D}$ satisfies the condition (G,9), $\mathcal{D}$ has an invariant connection. Moreover, a local expression of the connection satisfies the inequalities (I) and (II) in I.6. However, this case is rather trivial, because the geodesics starting from the identity e are given by $\xi(tu)$.

An invariant connection on the group of all volume preserving diffeomorphisms becomes the Euler's equation of perfect fluid. (Cf. [10, 11, 12] and [32].) In [32], the author proved that the natural right invariant connection on the group of all volume preserving diffeomorphisms satisfies the inequalities (I) and (II) and hence 1.6.2 Theorem holds.

II.2 A group of diffeomorphisms which commute with a compact group.

Let K be a compact subgroup of $\mathcal{D}$. Then, by Corollary [24 p202] and Theorem [24 p208], K is a compact Lie group acting smoothly on M. Thus, there is a smooth riemannian metric on M by which every element $g \in K$ becomes an isometry.

Let $\mathcal{D}_K = \{ \varphi \in \mathcal{D} : k\varphi = \varphi k$ for every $k \in K \}$.

The goal of this section is to prove the following :

2.2.1 Theorem $\mathcal{D}_K$ <u>is a strong ILB- and a strong ILH- Lie subgroup of</u> $\mathcal{D}$, <u>where</u> $\mathcal{D}$ <u>is regarded respectively as a strong ILB-Lie group modeled on</u> $\{ \Gamma(T_M), \gamma^k(T_M), k \in N(1) \}$ <u>and a strong ILH-Lie group modeled on</u> $\{ \Gamma(T_M), \Gamma^k(T_M), k \in N(\dim M + 5) \}$.

Proof. Let Exp be the exponential mapping with respect to the riemannian metric mentioned above. Let ξ be the mapping defined by $\xi(u)(x) = \mathrm{Exp}_x u(x)$. Thus, we see $g\mathrm{Exp}\, u(x) = \mathrm{Exp}\, dg(u(x))$ for any $u \in \Gamma(T_M)$, and hence $g\xi(u) = \xi(u)g$ if and only if $dgu(g^{-1}(x)) = u(x)$ for a sufficiently small u.

Let $A : \Gamma(T_M) \longmapsto \Gamma(T_M)$ be the linear mapping defined by $(Au)(x) = \int_K dgu(g^{-1}(x))dg$, where dg is an invariant measure of K. A can be extended to a bounded linear operator $A : \Gamma^k(T_M) \longmapsto \Gamma^k(T_M)$ for every $k \in N(d)$, $d = \dim M + 5$, and to a bounded linear operator $A : \gamma^k(T_M) \longmapsto \gamma^k(T_M)$ for every $k \in N(1)$.

We define subspaces as follows :

$$\Gamma_K(T_M) = \{ u \in \Gamma(T_M) : dgu(g^{-1}(x)) = u(x) \text{ for every } g \in K \},$$

$$\tilde{\Gamma}_K(T_M) = \{ u \in \Gamma(T_M) : Au = 0 \}.$$

Let $\Gamma^k_K(T_M)$, $\tilde{\Gamma}^k_K(T_M)$ be the closures of $\Gamma_K(T_M)$, $\tilde{\Gamma}_K(T_M)$ in $\Gamma^k(T_M)$ respectively and $\gamma^k_K(T_M)$, $\tilde{\gamma}^k_K(T_M)$ be the closures of $\Gamma_K(T_M)$, $\tilde{\Gamma}_K(T_M)$ in $\gamma^k(T_M)$ respectively. Since $u = Au + (u - Au)$, $Au \in \Gamma_K(T_M)$, $u - Au \in \tilde{\Gamma}_K(T_M)$ for any $u \in \Gamma(T_M)$, we see that $\Gamma(T_M) = \Gamma_K(T_M) \oplus \tilde{\Gamma}_K(T_M)$, $\Gamma^k(T_M) = \Gamma^k_K(T_M) \oplus \tilde{\Gamma}^k_K(T_M)$ and $\gamma^k(T_M) = \gamma^k_K(T_M) \oplus \tilde{\gamma}^k_K(T_M)$.

Obviously, $\xi : \mathcal{U} \cap \Gamma_K(T_M) \longmapsto \mathcal{D}_K$ is an into-homeomorphism, and hence this gives a strong ILB- and a strong ILH- Lie group structures on $\mathcal{D}_K$. By definition,

$\mathcal{D}_K$ is a strong ILB- and a strong ILH- Lie subgroup of $\mathcal{D}$.

Remark. It is easy to see that $A : \Gamma(T_M) \mapsto \Gamma(T_M)$ satisfies the inequality

$$\|Au\|_k \leq C\|u\|_k + D_k\|u\|_{k-1} ,$$

where $k \geq \dim M + 5$ if the norm is H-norm, and $k \geq 1$ if the norm is B-norm. Hence, we see $\|u - Au\|_k \leq C'\|u\|_k + D_k\|u\|_{k-1}$. Therefore

$$\|Au\|_k + \|u - Au\|_k \leq C''\|u\|_k + D''_k(\|u - Au\|_{k-1} + \|Au\|_{k-1}).$$

So, the above splitting is an ILB- and ILH- normal splitting.

II.3 Fibre preserving diffeomorphisms.

Let $\mathcal{F}$ be a smooth fibering of M with the base spase N and the projection π. N is a closed C^∞-manifold. The tangent space of the fibres define a smooth involutive subbundle F of T_M. Let E be a complementary subbundle of F, i.e. $T_M = F \oplus E$ (Whitney sum). F_x means the fibre of F at $x \in M$ and $\mathcal{F}_y$ means the fibre of $\mathcal{F}$ at $y \in N$. There exists a smooth connection on M such that $\operatorname{Exp} X \in \mathcal{F}_{\pi x}$ if $X \in F_x$ for any $x \in M$.

On the other hand, since $d\pi : E_x \mapsto T_{\pi x}N$ (the tangent space at πx) is an isomorphism, for any curve $c(t)$ in N such that $c(0) = \pi x$, there is a lift $\tilde{c}(t)$ such that $\tilde{c}(0) = x$, $\frac{d}{dt}\tilde{c}(t) \in E_{c(t)}$ and $\pi\tilde{c}(t) = c(t)$. Let Exp' be an exponential mapping defined by a smooth connection on N, and $\tilde{E}(tX)$ the lift of $\operatorname{Exp} td\pi X$, $t \in [0,1]$, for any $X \in E$. For any $X \in F_x$, $Y \in E_x$, we define $\tilde{E}(X,Y)$ by $\tilde{E}((d\pi_{\operatorname{Exp}X})^{-1}d\pi Y)$, where $d\pi_z^{-1}$ is the inverse of $d\pi : E_z \mapsto T_{\pi z}N$. The derivatives of $\tilde{E}$ at $X = 0$, $Y = 0$ is obviously the identity. Since M is compact, there are relatively compact, open tubular neighborhoods W_F, W_E of zero sections of F, E respectively such that $\tilde{E} : \overline{W}_{F,x} \oplus \overline{W}_{E,x} \longmapsto M$ is an into-diffeomorphism for every $x \in M$, where $W_{F,x} = W_F \cap F_x$ etc. and $-$ means the closure.

We define a mapping ξ' of $\Gamma(W_F) \oplus \Gamma(W_E)$ into the space of C^∞-mappings of M

into M by $\xi'(u+v)(x) = \widetilde{E}(u(x),v(x))$. Hence, if we restrict this mapping to small neighborhoods of zeros in C^1-topology, then ξ' is a homeomorphism into $\mathcal{D}$. Let ξ be the ILB- or ILH- coordinate mapping as in 2.1.1 Lemma. Obviously, $\xi^{-1}\xi'$ is defined from a smooth mapping $\mathrm{Exp}_{\pi X}^{-1}\widetilde{E}((d\pi_{\mathrm{Exp}X})^{-1}d\pi Y)$ and $d(\xi^{-1}\xi')_0$ = identity.

2.3.1 Lemma <u>There are neighborhoods</u> V_F, V_E <u>of zeros in</u> $\Gamma^d(W_F)$, $\Gamma^d(W_E)$ (resp. $\gamma^1(W_F)$, $\gamma^1(W_E)$) <u>respectively such that</u> $(V_F \oplus V_E , \xi')$ <u>is an ILH-</u> (resp. ILB-) <u>coordinate of</u> $\mathcal{D}$ <u>at the identity, where</u> $d = \dim M + 5$.

The above lemma is not proved here. We will give some inequalities in the section II.5 and the inverse function theorem in § III. The above lemma is an immediate conclusion of the inverse function theorem. Here, we will assume the above lemma.

Let $\Gamma_{\mathfrak{F}}(E)$ be the totality of $v \in \Gamma(E)$ such that $d\pi v$ is constant along the fibres, that is, $d\pi v(x') = d\pi v(x)$ if x, x' are contained in same fibre of $\mathfrak{F}$. For $u \in V_F$, $v \in V_E$, it is easy to see that $\xi'(u+v) \in \mathcal{D}_{\mathfrak{F}}$ if and only if $v \in V_E \cap \Gamma_{\mathfrak{F}}(E)$. Let $\Gamma^k_{\mathfrak{F}}(E)$ (resp. $\gamma^k_{\mathfrak{F}}(E)$) be the closure of $\Gamma_{\mathfrak{F}}(E)$ in $\Gamma^k(E)$ (resp. $\gamma^k(E)$.) The local coordinate of the fibre preserving diffeomorphisms $\mathcal{D}_{\mathfrak{F}}$ is given by the mapping $\xi' : V_F \cap \Gamma(F) \oplus V_E \cap \Gamma_{\mathfrak{F}}(E) \longmapsto \mathcal{D}_{\mathfrak{F}}$. It is not hard to check the conditions (N,1) - (N,7) by using the properties of $\mathcal{D}$, hence $(V_F \oplus V_E \cap \Gamma^d_{\mathfrak{F}}(E)$, ξ') is an ILH-coordinate of $\mathcal{D}_{\mathfrak{F}}$ and $(V_F \oplus V_E \cap \gamma^1_{\mathfrak{F}}(E), \xi')$ is an ILB- coordinate of $\mathcal{D}_{\mathfrak{F}}$.

There is also a splitting of $\Gamma(T_M)$. Let dV_F be a smooth volume element along the fibres such that $\int_{\mathfrak{F}_y} dV_F \equiv 1$ for any $y \in N$. Let $\widetilde{\Gamma}_{\mathfrak{F}}(E)$ be the space defined by $\{ u \in \Gamma(E) : \int_{\mathfrak{F}_y} d\pi u(x)\, dV_F(x) \equiv 0$ for every $y \in N \}$, $\widetilde{\Gamma}^k_{\mathfrak{F}}(E)$ (resp. $\widetilde{\gamma}^k_{\mathfrak{F}}(E)$) the closure of $\widetilde{\Gamma}_{\mathfrak{F}}(E)$ in $\Gamma^k(E)$ (resp. $\gamma^k(E)$.) Then, it is easy to show the following splitting :

$\Gamma(T_M) = (\Gamma(F) \oplus \Gamma_{\mathfrak{F}}(E)) \oplus \widetilde{\Gamma}_{\mathfrak{F}}(E)$, $\quad \Gamma^k(T_M) = (\Gamma^k(F) \oplus \Gamma^k_{\mathfrak{F}}(E)) \oplus \widetilde{\Gamma}^k_{\mathfrak{F}}(E)$ and

$\gamma^k(T_M) = (\gamma^k(F) \oplus \gamma^k_{\mathfrak{F}}(E)) \oplus \widetilde{\gamma}^k_{\mathfrak{F}}(E)$.

What we proved is the following :

2.3.2 Theorem <u>The fibre preserving diffeomorphisms $\mathcal{D}_{\mathcal{F}}$ is a strong ILB- and a strong ILH- Lie subgroup of</u> $\mathcal{D}$.

<u>Remark 1</u> The above splitting $\Gamma(T_M) = (\Gamma(F) \oplus \Gamma_{\mathcal{F}}(E)) \oplus \tilde{\Gamma}_{\mathcal{F}}(E)$ is both ILB- and ILH- normal splitting.

<u>Remark 2</u> Let $\mathcal{D}_{[\mathcal{F}]}$ be the subgroup of $\mathcal{D}_{\mathcal{F}}$ such that every element of $\mathcal{D}_{[\mathcal{F}]}$ leaves each fibre invariant. Then, $\mathcal{D}_{[\mathcal{F}]}$ is a closed and normal subgroup of $\mathcal{D}_{\mathcal{F}}$. In §III, it will be proved that $\mathcal{D}_{[\mathcal{F}]}$ is a strong ILB- and a strong ILH- Lie subgroup of $\mathcal{D}_{\mathcal{F}}$. The factor set $\mathcal{D}_{[\mathcal{F}]} \backslash \mathcal{D}_{\mathcal{F}}$ is an open subgroup of $\mathcal{D}(N)$.

<u>Remark 3</u> By a similar method, we can prove the following :

2.3.3 Theorem <u>The group $\mathcal{D}_E$ of all automorphisms of vector bundle E over M has a strong ILB- and a strong ILH- Lie group structures. The subgroup $\mathcal{D}_{[E]}$ of all elements which leave each fibre invariant is a closed and normal subgoup of $\mathcal{D}_E$ and a strong ILB- and a strong ILH- Lie subgroup of</u> $\mathcal{D}_E$.

II.4. A group of diffeomorphisms which leaves a submanifold invariant.

Let S be a closed submanifold of M without boundary. Since M is compact, S is compact and hence there is a smooth connection on M such that S is a totally geodesic submanifold of M.

Let $\mathcal{U}$ be the same neighborhood of 0 in $\Gamma^1(T_M)$ as in II.1 and let $\xi(u)(x) = \mathrm{Exp}_x u(x)$. We put $\Gamma_S(T_M) = \{ u \in \Gamma(T_M) : u(x) \in T_x S$ for any $x \in S \}$, where $T_x S$ is the tangent space of S at x. Put $\Gamma_{[S]}(T_M) = \{ u \in \Gamma(T_M) : u(x) = 0$ for every $x \in S \}$.

Let $\mathcal{D}(M,S) = \{ \varphi \in \mathcal{D}(M) : \varphi(S) = S \}$,

$$\mathcal{D}(M,[S]) = \{ \varphi \in \mathcal{D}(M) : \varphi(x) = x \text{ for any } x \in S \}.$$

It is not hard to see that $\xi : \mathfrak{U} \cap \Gamma_S(T_M) \mapsto \mathcal{D}(M,S)$, $\xi : \mathfrak{U} \cap \Gamma_{[S]}(T_M) \mapsto \mathcal{D}(M,[S])$ are into homeomorphisms, if we take $\mathfrak{U}$ very small in $\gamma^1(T_M)$.

Let $\gamma_S^k(T_M)$, $\gamma_{[S]}^k(T_M)$ be the closures of $\Gamma_S(T_M)$, $\Gamma_{[S]}(T_M)$ in $\gamma^k(T_M)$ respectively.

Denote by $r_S u$ the restriction of a vector field u onto S. Let ν_S be the normal bundle of S and π the projection of it. $d\pi$ is the derivative of π. It is also easy to prove that $r_S : \Gamma_S(T_M) \mapsto \Gamma(T_S)$, $d\pi r_S : \Gamma(T_M) \mapsto \Gamma(T_S)$ can be extended to bounded linear mappings $r_S : \gamma_S^k(T_M) \mapsto \gamma^k(T_S)$, $d\pi r_S : \gamma^k(T_M) \mapsto \gamma^k(T_S)$. (Remark that r_S (resp. $d\pi r_S$) can <u>not</u> be extended to a bounded linear mapping of $\Gamma_S^k(T_M)$ into $\Gamma^k(T_S)$ (resp. of $\Gamma^k(T_M)$ into $\Gamma^k(T_S)$).

2.4.1 Lemma <u>There exist linear imbeddings</u> $i : \Gamma(T_S) \mapsto \Gamma_S(T_M)$, $j : \Gamma(\nu_S) \mapsto \Gamma(T_M)$ <u>such that</u> $r_S i =$ identity <u>and</u> $r_S j =$ identity <u>and that</u> i <u>and</u> j <u>can be extended to bounded linear imbeddings</u> $i : \gamma^k(T_S) \mapsto \gamma_S^k(T_M)$, $j : \gamma^k(\nu_S) \mapsto \gamma^k(T_M)$ <u>respectively.</u>

<u>Moreover,</u> $u = (u - ir_S u) + ir_S u$, $v = (v - j(1-d\pi)r_S v) + j(1-d\pi)r_S v$ <u>give ILB-normal splitting of</u> $\Gamma_S(T_M)$ <u>and</u> $\Gamma(T_M)$. <u>Namely,</u> $\Gamma_S(T_M) = \Gamma_{[S]}(T_M) \oplus i\Gamma(T_S)$ <u>and</u> $\Gamma(T_M) = \Gamma_S(T_M) \oplus j\Gamma(\nu_S)$ <u>are ILB-normal splittings.</u>

Proof. S has a tubular neighborhood which is diffeomorphic to a neighborhood of zero section of ν_S. Since ν_S has a riemannian connection, any vector field $u \in \Gamma(T_S)$ (resp. any section $u \in \Gamma(\nu_S)$) can be extended to a vector field on ν_S as a horizontal (resp. as a vertical) vector field. Therefore, u can be extended to a smooth vector field $\tilde{u}$ defined on the tubular neighborhood of S. Let f be a smooth function such that $f|S \equiv 1$ and the support of f is contained in the tubular neighborhood. $f\tilde{u}$ can be regarded as a smooth vector field on M. We define $i(u) = f\tilde{u}$ and $j(u) = f\tilde{u}$. All others are easy to prove from this definition.

Since $\xi : \mathfrak{U} \cap \Gamma_S(T_M) \mapsto \mathcal{D}(M,S)$ and $\xi : \mathfrak{U} \cap \Gamma_{[S]}(T_M) \mapsto \mathcal{D}(M,[S])$ are into-homeomorphisms, we have the following :

2.4.2 Theorem $\mathcal{D}(M,S)$ <u>is a strong ILB-Lie subgroup of</u> $\mathcal{D}$ <u>and</u> $\mathcal{D}(M,[S])$ <u>is a strong ILB-Lie subgroup of</u> $\mathcal{D}(M,S)$. $\mathcal{D}(M,[S])$ <u>is a normal subgroup of</u> $\mathcal{D}(M,S)$ <u>and the factor group</u> $\mathcal{D}(M,[S])\backslash\mathcal{D}(M,S)$ <u>is an open subgroup of</u> $\mathcal{D}(S)$.

Now, suppose S is a submanifold with boundary. Then, it seems to be impossible to find an ILB-splitting of $\Gamma_S(T_M)$ in $\Gamma(T_M)$ or of $\Gamma_{[S]}(T_M)$ in $\Gamma_S(T_M)$. It might be related to the following fact : Consider the space of all C^∞-functions Γ on the interval $[-1,1]$. Let Γ_o be the totality of functions f such that f and the derivatives of f vanish at the origin 0. Then, there is no ILB-splitting of Γ_o in Γ, because if we take the closure of Γ_o in γ^k (the space of all C^k-functions with C^k-topology) then γ^k_o = the closure of Γ_o is the totality of C^k-functions f such that f and the derivative of f up to k-th order vanish at 0. Thus, we have $\gamma^k_o \cap \Gamma \supsetneq \Gamma_o$.

So, if we want to extend the above theorem, we have to relax the condition of strong ILB-Lie groups.

2.4.3 Definition. A subgroup H of a strong ILB- (resp. ILH-) Lie group G modeled on $\{\mathbb{E}, E^k, k \in N(d)\}$ will be called <u>a strong ILB- (resp. ILH-) Lie subgroup without splitting condition,</u> if the following are satisfied :

(a) There is a closed subspace $\mathbb{F} \subset \mathbb{E}$ and an integer $d' \geq d$ such that $F^k \cap \mathbb{E} = \mathbb{F}$ for any $k \geq d'$, where F^k is the closure of $\mathbb{F}$ in E^k.

(b) Regarding G as a strong ILB- (resp. ILH-) Lie group modeled on $\{\mathbb{E}, E^k, k \in N(d')\}$, there is an ILB-(resp. ILH-) coordinate (U, ξ) such that $\xi(U \cap \mathbb{F}) = \tilde{U}_o(H)$, where $\tilde{U} = \xi(U \cap H)$. (U is an open neighborhood of 0 in $E^{d'}$.)

<u>Remark 1</u> If H is a strong ILB- (resp. ILH-) Lie subgroup without splitting condition of a strong ILB- (resp. ILH-) Lie group G, then H itself is a strong ILB- (resp. ILH-) Lie group.

<u>Remark 2</u> Let $\mathcal{D}_{[x]}(M)$ is the totality of φ in $\mathcal{D}(M)$ such that $(d\varphi)_x = id$.

and $(d^k\varphi)_x = 0$ for every $k > 1$. Then, $\mathcal{D}_{[x]}(M)$ is <u>not</u> a strong ILB-Lie subgroup of $\mathcal{D}(M)$ without splitting condition.

Remark 3 Recall the proof of the previous theorem. This theorem is still true for the subset S which satisfies the following conditions, if we do not care about the splitting condition.

(a) S is closed in M.

(b) S is a disjoint union of finitely many submanifolds S_i, $i = 1 \sim m$ such that if $\overline{S}_i \cap S_j \neq \emptyset$, then $\overline{S}_i \supset S_j$.

(c) There is a smooth riemannian metric on M such that for any $x \in M$, there is an open convex neighborhood U_x such that $U_x \cap S_i$ is also a convex subset for any S_i.

The condition (c) is very strong and technical. For instance, because of the condition (c), we must exclude the subset like Fig. 1 or 2.

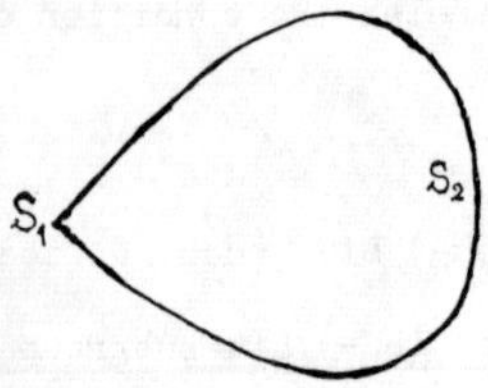

Fig. 1

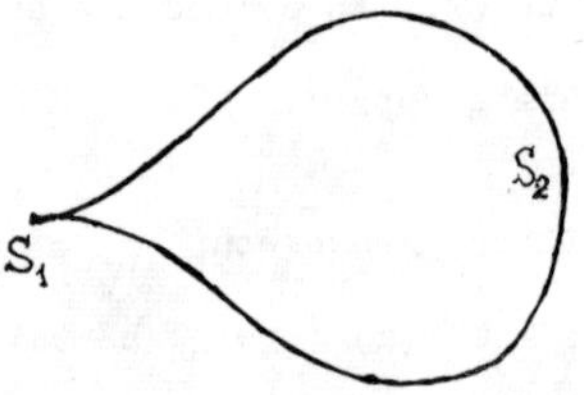

Fig. 2

However, it is easy to show that $\mathcal{D}(M,S)$, where S = Fig. 1 in a 2-dimensional torus, is a strong ILB- Lie subgroup of $\mathcal{D}(M)$ without splitting condition and this may be true for S = Fig. 2.

At least at this moment, the author does not know what is the weakest conditions for S under which $\mathcal{D}(M,S)$ is a strong ILB-Lie group or a strong ILB-Lie subgroup of $\mathcal{D}(M)$ without splitting condition. Such conditions, if we could know it, might be very powerfull for the study of singularlities of mappings.

II.5 Some inequalities.

All inequalities given here are special case of Theorem A in [30] or can be proved easily by the same mathod. They are however much easier to use, but the tiresome computations will not be repeated here. They can be also proved directly by careful computations together with 2.1.3 Lemma.

Let E, F be smooth finite dimensional vector bundle over M.

2.5.1 Lemma <u>Let</u> α <u>be an element of</u> $\Gamma(E^* \otimes F)$ <u>and</u> $u \in \Gamma(E)$. <u>Then, we have</u>

$$\|\alpha u\|_k \leq C\{\|\alpha\|_k\|u\|_{k_0} + \|\alpha\|_{k_0}\|u\|_k\} + D_k\|\alpha\|_{k-1}\|u\|_{k-1}$$

<u>for every</u> $k \geq \dim M + 5$, <u>if the norm is H-norm and</u> $k \geq 1$ <u>if the norm is B-norm,</u> C, D_k <u>are positive constants and</u> C <u>does not depend on</u> k, $k_0 = [\frac{1}{2}\dim M] + 1$ ($\| \ \|_k$ = H-norm) or $k_0 = 0$ ($\| \ \|_k$ = B-norm).

<u>Remark</u> It can be shown by a direct computation that the above inequality holds in fact for $k \geq \dim M + 3$ for H-norms.

Let $u \otimes v$ denote the tensor product $(u \otimes v)(x) = u(x) \otimes v(x)$.

2.5.2 Lemma <u>Let</u> k_0, C, D_k <u>be as above. If</u> $k \geq \dim M + 5$ (H-norm) <u>or</u> $k \geq 1$ (B-norm), <u>then</u>

$$\|u \otimes v\|_k \leq C\{\|u\|_k\|v\|_{k_0} + \|u\|_{k_0}\|v\|_k\} + D_k\|u\|_{k-1}\|v\|_{k-1} .$$

<u>Remark</u> In fact, the above inequality holds for $k \geq \dim M + 3$ for H-norms.

We now back to the situation of 2.1.3 Lemma and use the same notations. There $\Phi : \Gamma(W) \mapsto \Gamma(F)$ was defined from the smooth fibre preserving mapping φ.

Let $(\partial^r\varphi)_Y(X_1,\ldots,X_r)$ be the r-th derivative of φ along the fibres. $\partial^r\varphi$ is a smooth fibre preserving mapping of W into $(E^* \otimes \ldots \otimes E^*) \otimes F$ and the r-th derivative $d^r\Phi$ of Φ is given by

$$(d^r\Phi)_u(v_1,\ldots,v_r)(x) = (\partial^r\varphi)_{u(x)}(v_1(x),\ldots,v_r(x)).$$

Apply 2.5.1 to $\partial^r\varphi$ and use 2.5.2 successively. Apply 2.1.3 to the resulting inequality. Then, we have the following :

2.5.3 Lemma Let $k_1 = \dim M + 5$, if the norm is H-norm and let $k_1 = 1$, if the norm is B-norm. If u is restricted in a bounded set in $\Gamma^{k_1}(W)$ (H-norm) or $\gamma^{k_1}(W)$ (B-norm), then the following inequality holds for $k \geq k_1 + 1$:

$$\|(d^r\Phi)_u(v_1,\dots,v_r)\|_k \leq C\{\|u\|_k\|v_1\|_{k_1}\cdots\|v_r\|_{k_1} + \sum_{j=1}^{r}\|v_1\|_{k_1}\cdots\|v\|_{j-1}\|_{k_1}\|v_j\|_k\|v_{j+1}\|_{k_1}\cdots\|v_r\|_{k_1}\}$$

$$+ P_k(\|u\|_{k-1})\|v_1\|_{k-1}\cdots\|v_r\|_{k-1} ,$$

where C is a positive constant independent from k and P_k is a polynomial with positive coefficients depending on k.

Remark $\xi^{-1}\xi'$ in 2.3.1 Lemma satisfies the same inequality as above, because this is defined from a smooth fibre preserving mapping.

§ III Basic theorems I

III.1 The inverse function theorem.

The goal here is to prove 3.1.1 Theorem below. 2.3.1 Lemma is an immediate conclusion of 2.5.3 Lemma and the result here.

Let $\{ \mathbb{E}, E^k, k \in N(d) \}$ be a Sobolev chain.

3.1.1 Theorem (inverse function theorem) Let U, U' be open neighborhood of 0 in E^d. Suppose a mapping $\Phi : U \cap \mathbb{E} \mapsto U' \cap \mathbb{E}$ with $\Phi(0) = 0$ satisfies the following :

(1) Φ can be extended to a C^∞-mapping of $U \cap E^k$ into $U' \cap E^k$ for every $k \in N(d)$.

(2) $(d\Phi)_0 : E^k \mapsto E^k$ is an isomorphism for every $k \in N(d)$.

(3) For every $u \in U \cap \mathbb{E}$, $v \in \mathbb{E}$,

$$\|(d\Phi)_u v\|_k \leq C\{\|u\|_k \|v\|_d + \|v\|_k\} + P_k(\|u\|_{k-1})\|v\|_{k-1}, \quad k \geq d+1,$$

$$\|(d^2\Phi)_u(v_1, v_2)\|_k \leq C\{\|u\|_k \|v_1\|_d \|v_2\|_d + \|v_1\|_k \|v_2\|_d + \|v_1\|_d \|v_2\|_k\} + P_k(\|u\|_{k-1})\|v_1\|_{k-1}\|v_2\|_{k-1}, \quad k \geq d+1,$$

(4) $\|(d\Phi)_0 v\|_k \geq C'\|v\|_k - D_k\|v\|_{k-1}, \quad k \geq d+1,$

where C, C', D_k are positive constants, and C, C' are independent from k and P_k is a polynomial with positive coefficients depending on k.

Then, there are neighborhoods W, W' of 0 in E^d such that $\Phi : W \cap E^k \mapsto W' \cap E^k$ is a C^∞-diffeomorphism for every $k \in N(d)$. Moreover, Φ^{-1} satisfies the same inequalities as (3) and (4), that is, Φ^{-1} has the same properties as Φ.

The proof will be given in the several lemmas below.

By the condition (4), we see $\|v\|_k \geq C'\|(d\Phi)_0^{-1}v\|_k - D_k'\|(d\Phi)_0^{-1}v\|_{k-1}$, hence $\|(d\Phi)_0^{-1}v\|_k \leq C'^{-1}\|v\|_k + D_k''\|v\|_{k-1}$. Therefore, $(d\Phi)_0^{-1}\Phi : U_1 \cap \mathbb{E} \mapsto U' \cap \mathbb{E}$ satisfies the conditions (1) ~ (3) for a small open neighborhood U_1 of 0 in E^d. The condition (4) is trivial in this case. Thus, we assume, henceforth, that $(d\Phi)_0 = id.$. There is an open and convex neighborhood W of 0 in E^d such that Φ is a C^∞-

diffeomorphism of $\overline{W}$ onto an open neighborhood $\overline{W}'$ of $0 \in E^d$ and that $\overline{W}$ is contained in $1/8C$ -neighborhood of 0 in E^d.

Since $((d\Phi)_y - I)v = \int_0^1 (d^2\Phi)_{ty}(y,v)dt$ (where I = identity), we have the following inequality using the condition (3) :

$$\text{(a)} \quad \|(d\Phi)_y v - v\|_k \leqslant \frac{C}{2}\|y\|_k\|y\|_d\|v\|_d + C\{\|y\|_d\|v\|_k + \|y\|_k\|v\|_d\}$$
$$+ P_k'(\|y\|_{k-1})\|y\|_{k-1}\|v\|_{k-1} \, .$$

3.1.2 Lemma <u>Notations and assumptions being as above,</u> $(d\Phi)_u : E^k \mapsto E^k$ <u>is an isomorphism for any</u> $u \in W \cap E^k$ <u>and for any</u> $k \in N(d)$.

Proof. Let W^k be the totality of $u \in W \cap E^k$ such that $(d\Phi)_u : E^k \mapsto E^k$ is an isomorphism. Then, $W^d = W$ and W^k is a non-empty open subset of $W \cap E^k$. We will prove $W^k = W \cap E^k$ by induction. So, assume $W^s = W \cap E^s$ for $d \leqslant s \leqslant k-1$. We have only to show that W^k is closed in the connected set $W \cap E^k$.

Let z be a boundary point of W^k in $W \cap E^k$. Apply the innequality (a) to $(d\Phi)_z v = v + (d\Phi)_z v - v$ and we get

$$\|(d\Phi)_z v\|_k \geqslant \frac{7}{8}\|v\|_k - P_k''(\|z\|_k)\|v\|_{k-1} \, .$$

Since $(d\Phi)_z : E^s \mapsto E^s$ is an isomorphism for $d \leqslant s \leqslant k-1$, the above inequality shows the closedness of the image $(d\Phi)_z E^k$.

Let $\{z_n\}$ be a sequence in W^k converging to z. Let v be an arbitrary element in E^k. Since $z_n \in W^k$, there is $u_n \in E^k$ such that $(d\Phi)_{z_n} u_n = v$. We see easily that

$$\|v\|_k \geqslant \frac{7}{8}\|u_n\|_k - D_k'\|u_n\|_{k-1} - |||(d\Phi)_{z_n} - (d\Phi)_z|||\cdot\|u_n\|_k \, ,$$

where $|||\ \ |||$ means the operator norm. Hence, for sufficiently large n, we have

$$\|v\|_k \geqslant \frac{1}{2}\|u_n\|_k - D_k'\|u_n\|_{k-1} .$$

Since $\|u_n\|_{k-1}$ is bounded by the assumption of induction, we see that $\|u_n\|_k$ is bounded. Therefore, the inequality $\| v - (d\Phi)_z u_n \|_k \leqslant |||(d\Phi)_{z_n} - (d\Phi)_z|||\|u_n\|_k$ implies that $v \in (d\Phi)_z E^k$.

Thus, $(d\Phi)_z : E^k \mapsto E^k$ is surjective, hence an isomorphism because $(d\Phi)_z : E^d \mapsto E^d$ is an isomorphism. Therefore, W^k is closed in $W \cap E^k$ and hence $W^k = W \cap E^k$.

By virtue of the above lemma, we see that $\Phi(W \cap E^k)$ is an open subset of $W' \cap E^k$ for every $k \in N(d)$ and $\Phi : W \cap E^k \mapsto \Phi(W \cap E^k)$ is a C^∞-diffeomorphism.

3.1.3 Lemma <u>Notations and assumptions being as above,</u> we have $\Phi(W \cap E^k) = W' \cap E^k$.

Proof. This is proved by induction. We assume $\Phi(W \cap E^s) = W' \cap E^s$ for $d \leq s \leq k-1$. Apply the inequality (a) to $\Phi(y) = \Phi(y) - \Phi(0) = y + \int_0^1 ((d\Phi)_{ty}y - y)dt$, and we have

$$\text{(b)} \quad \|\Phi(y)\|_k \geq \frac{5}{8}\|y\|_k - P_k'(\|y\|_{k-1})\|y\|_{k-1}^2 .$$

Let z be a boundary point of $\Phi(W \cap E^k)$ in $W' \cap E^k$, and let $\{z_n\}$ be a sequence in $\Phi(W \cap E^k)$ converging to z. There is x_n for every n such that $\Phi(x_n) = z_n$. $\{x_n\}$ is bounded in the norm $\| \ \|_{k-1}$. Therefore, using the above inequality (b), we see that $\|x_n\|_k$ is bounded. Apply (a) to

$$z_n - z_m = x_n - x_m + \int_0^1 ((d\Phi)_{x_m + t(x_n - x_m)}(x_n - x_m) - (x_n - x_m))\, dt,$$

and we get

$$\begin{aligned}\|z_n - z_m\|_k \geq \frac{3}{4}\|x_n - x_m\|_k &- C(\|x_n\|_k + \|x_m\|_k)\|x_n - x_m\|_d \\ &- C(\|x_n\|_k + \|x_m\|_k)\|x_n - x_m\|_k\|x_n - x_m\|_d \\ &- P_k(\|x_n\|_{k-1} + \|x_m\|_{k-1})\|x_n - x_m\|_{k-1} .\end{aligned}$$

Since $\|x_n\|_k$ is bounded, we see for sufficiently large n, m,

$$\|z_n - z_m\|_k \geq \frac{1}{2}\|x_n - x_m\|_k - D_k'\|x_n - x_m\|_{k-1} .$$

Notice that $\{x_n\}$ is a Cauchy sequence in the norm $\| \ \|_{k-1}$. Therefore, the above inequality shows that $\{x_n\}$ is a Cauchy sequence in the norm $\| \ \|_k$. Thus, $\{x_n\}$ converges to annelement $x_0 \in \overline{W} \cap E^k$ and $\Phi(x_0) = z$. However, since $\Phi(W \cap E^{k-1}) =$

$W' \cap E^{k-1}$, there is $x \in W \cap E^{k-1}$ such that $\Phi(x) = z$. This must be equal with x_o, hence $x_o \in W \cap E^k$. Therefore $\Phi(W \cap E^k)$ is closed in $W' \cap E^k$. Thus, $\Phi(W \cap E^k) = W' \cap E^k$ because $W' \cap E^k$ is connected.

Now, to complete the proof of 3.1.1, we have only to show the following :

3.1.4 Lemma <u>Notations and assumptions being as above, there is a neighborhood</u> W_1' <u>of</u> 0 <u>in</u> E^d <u>such that the following inequalities hold for any</u> $u \in W_1' \cap E^k$, $k \in N(d)$.

$$\|(d\Phi^{-1})_u v\|_k \leq C'\{\|u\|_k\|v\|_d + \|v\|_k\} + P_k'(\|u\|_{k-1})\|v\|_{k-1} ,$$

$$\|(d^2\Phi^{-1})_u(v_1,v_2)\|_k \leq C'\{\|u\|_k\|v_1\|_d\|v_2\|_d + \|v_1\|_k\|v_2\|_d + \|v_1\|_d\|v_2\|_k\}$$
$$+ P_k'(\|u\|_{k-1})\|v_1\|_{k-1}\|v_2\|_{k-1} ,$$

$$\|(d\Phi^{-1})_o v\|_k \geq C''\|v\|_k - D_k''\|v\|_{k-1} .$$

Proof. Since $\|(d\Phi)_o v\|_k \leq C\|v\|_k + D_k\|v\|_{k-1}$, $\|(d\Phi)_o v\|_k \geq C'\|v\|_k - D_k'\|v\|_{k-1}$, the last inequality is easy to get, hence we have only to prove the inequalities for $(d\Phi)_o\Phi^{-1}$. Thus, we may assume $(d\Phi)_o = (d\Phi^{-1})_o = I$, and we can use the inequalities obtained in the proof of the above lemmas.

There is a neighborhood W_1' of 0 in E^d such that $W_1' \subset W'$, and $\|\Phi^{-1}(u)\|_d \leq C_1$, $\|(d\Phi^{-1})_u w\|_d \leq C_1\|w\|_d$ for every $u \in W_1'$. Use the inequality (b), and we have

$$\|w\|_k \geq \frac{5}{8}\|\Phi^{-1}(w)\|_k - P_k'(\|\Phi^{-1}(w)\|_{k-1})\|\Phi^{-1}(w)\|_{k-1}^2 .$$

Compute successively and use $\|\Phi^{-1}(w)\|_d \leq C_1$. Then, $\|\Phi^{-1}(w)\|_k \leq \frac{5}{8}\|w\|_k + P_k''(\|w\|_{k-1})$.

Apply the inequality (a) to $(d\Phi)_z v = v + (d\Phi)_z v - v$, and we get

$$\|(d\Phi)_z v\|_k \geq \frac{7}{8}\|v\|_k - C\|z\|_k\|v\|_d - \frac{C}{2}\|z\|_k\|z\|_d\|v\|_d - P_k'(\|z\|_{k-1})\|z\|_{k-1}\|v\|_{k-1} ,$$

hence

$$\|w\|_k \geq \frac{7}{8}\|(d\Phi^{-1})_y w\|_k - C'\|\Phi^{-1}(y)\|_k\|w\|_d - P_k''(\|\Phi^{-1}(y)\|_{k-1})\|(d\Phi^{-1})_z w\|_{k-1} .$$

Compute successively and use $\|(d\Phi^{-1})_z w\|_d \leq C_1\|w\|_d$ and the above inequality for

$\Phi^{-1}(y)$. Then, we have

$$\|(d\Phi^{-1})_y w\|_k \leq C'\{\|y\|_k\|w\|_d + \|w\|_k\} + P_k(\|y\|_{k-1})\|w\|_{k-1}.$$

Apply the obtained inequalities to

$$(d^2\Phi^{-1})_y(v_1,v_2) = -(d\Phi^{-1})_y(d^2\Phi)_{\Phi^{-1}(y)}((d\Phi^{-1})_y v_1, (d\Phi^{-1})_y v_2)$$

and use the inequality (3) for $(d^2\Phi)$. Then, we get the desired inequality. We omit here the tiresome computation.

3.1.5 Definition. Henceforth, we use the following terminology. A mapping Φ of $U \cap \mathbb{E}$ into $U' \cap \mathbb{E}$ will be called a $C^\infty ILBC^2$-normal (or $C^\infty ILHC^2$-normal) mapping, if Φ satisfies the conditions (1) and (3) in 3.1.1 Theorem.

Remark It is not hard to see that the composition of $C^\infty ILBC^2$-normal mappings is also a $C^\infty ILBC^2$-normal mapping. In fact, this is used in the above proof. As it was shown above, the inverse function theorem holds in the category of $C^\infty ILBC^2$-normal mappings.

III.2 Applications

Now, we back to the situation of § II.2. K is a compact group in $\mathcal{D}$, (U,ξ) is an ILH- or ILB- coordinate at e and ξ is given by $\xi(u)(x) = \mathrm{Exp}_x u(x)$ by a smooth riemannian metric g_{ij} under which K becomes isometries. Recall the definition of the H-norm and the B-norm. Here, we use the connection ∇ and the volume element dV defined naturally by the riemannian metric g_{ij}. $<\nabla^s u, \nabla^s v>$ means also the inner product by that riemannian metric. We put $Ad(g)u(x) = dg\, u(g^{-1}(x))$ for any $g \in K$.

3.2.1 Lemma $\|Ad(g)u\|_k = \|u\|_k$ for any $g \in K$ with respect to both H-norm and B-norm.
Proof. We have only to show $<\nabla^s u, \nabla^s u>(g^{-1}x) = <\nabla^s Ad(g)u, \nabla^s Ad(g)u>(x)$ because $g^{-1*}dV = dV$ in the case of H-norm. Define a local coordinate at x and use the same local coordinate at $g^{-1}x$ through the isometry g^{-1}. Then, $Ad(g)u$ around x has the same local coordinate expression of u around $g^{-1}x$. Since g^{-1} is an isometry,

the riemannian metric around x and $g^{-1}x$ has the same local coordinate expression. Thus, we have the desired equality.

3.2.2 Theorem Let ξ be as above and let (U',ξ') be another ILH-(resp ILB-) coordinate at e of $\mathcal{D}$ such that $\xi^{-1}\xi'$ is a C^∞ILHC^2-normal (resp. C^∞ILBC^2-normal) mapping with $(d(\xi^{-1}\xi'))_0 = I$. Then, there is a neighborhood W of 0 in $\Gamma^d(T_M)$, $d = \dim M + 5$ (resp. in $\gamma^1(T_M)$) such that for any $g \in K$, $A'_g(u) = \xi'^{-1}(g\xi'(u)g^{-1})$ is a C^∞ILH (resp. ILB) C^2-normal mapping on $W \cap \Gamma(T_M)$. Moreover, the constants in the inequalities which appear in the definition of C^∞ILH (resp. ILB) C^2-normal mappings are common to all $g \in K$.

Proof. Notations being as above, let $A_g(u) = \xi^{-1}(g\xi(u)g^{-1})$. Then, $A'_g(u) = (\xi^{-1}\xi')^{-1}A_g(\xi^{-1}\xi')(u)$. By the inverse function theorem $(\xi^{-1}\xi')^{-1}$ is also a C^∞ILH (resp. ILB) C^2-normal mapping on a small neighborhood $W \cap \Gamma(T_M)$. Thus, we have only to show that A_g is a C^∞ILH (resp. ILB) C^2-normal mapping. Since $g \in K$ is an isometry, we see that $A_g(u) = Ad(g)u$ and $(d^2A_g)_u = 0$. Therefore, $\|(dA_g)_u v\|_k = \|Ad(g)v\|_k$ and $\|(d^2A_g)_u(v,w)\|_k = 0$. It is now clear that A_g is a C^∞ILH (resp. ILB) C^2-normal mapping and the constants in the inequalities are common to all $g \in K$.

Remark If $\xi^{-1}\xi'$ is defined from a smooth fibre preserving mapping, then $\xi^{-1}\xi'$ is a C^∞ILH (resp.ILB)C^2-normal mapping. (Cf. II.5.) Moreover, by the inverse function theorem, we see that strong ILB- or strong ILH- Lie group structures on $\mathcal{D}$, $\mathcal{D}_K$, $\mathcal{D}_{\mathcal{F}}$, $\mathcal{D}(M,S)$ etc. do not depend on the choice of connections, riemannian metrics or volume elements.

Suppose we have a strong ILB-Lie group G. Let $\mathfrak{g}^k$ be the tangent space of G^k at e and let $\mathfrak{g} = \cap\, \mathfrak{g}^k$ with the inverse limit topology. Let U be an open neighborhood of 0 in $\mathfrak{g}^d$ and (U,ξ) an ILB-coordinate of G at e. For a compact subgroup K of G, there is a neighborhood W of 0 in $\mathfrak{g}^d$ such that $h\xi(W)h^{-1} \subset \xi(U)$ for every $h \in K$. The following is a generalization of the theorem in II.2. (Cf. Theorem D in § 0.)

3.2.3 Theorem Assume $A_h(u) = \xi^{-1}(h\xi(u)h^{-1})$ is a $C^\infty ILBC^2$-normal mapping on $W \cap \mathfrak{g}$ for every $h \in K$ and the constants are common to all $h \in K$ in the inequalities which appear in the definition of $C^\infty ILBC^2$-normal mappings. Then, $G_K = \{ g \in G : gh = hg \text{ for any } h \in K \}$ is a strong ILB-Lie subgroup of G.

Proof. Since $\|(dA_{h^{-1}})_o v\|_k \leq C\|v\|_k + D_k\|v\|_{k-1}$, we see that $(dA_{h^{-1}})_o A_h$ is a C^∞-ILBC^2-normal mapping. We consider now the transformation R defined by

$$R(u) = \int_K (dA_{h^{-1}})_o A_h(u)\, dh,$$

where dh is an invariant measure on K such that $\int_K dh = 1$. It is easy to see that $(dR)_o = I$ and that R is a $C^\infty ILBC^2$-normal mapping. Thus, by the inverse function theorem, there is a neighborhood W_1 of 0 in $\mathfrak{g}^d$ such that $(W_1, \xi R^{-1})$ is an ILB-coordinate of G at e.

Remark that $(dA_g)_o\, v = \lim_{t \to 0} \frac{1}{t}\, g\xi(tv)g^{-1} = Ad(g)v$. Thus, we have

$$Ad(g)R(u) = \int_K (dA_{gh^{-1}})_o A_h(u)dh = \int_K (dA_{h^{-1}})_o A_h(A_g u)dh = R(A_g(u)).$$

Therefore, putting $\xi' = \xi R^{-1}$, we see $\xi'^{-1}(h\xi'(u)h^{-1}) = Ad(h)u$.

We define subspaces as follows

$$\mathfrak{g}_K = \{ u \in \mathfrak{g} : Ad(h) = u \text{ for all } h \in K \}$$

$$\tilde{\mathfrak{g}}_K = \{ u \in \mathfrak{g} : \int_K Ad(h)u\, dh = 0 \}.$$

Let $\mathfrak{g}_K^k$, $\tilde{\mathfrak{g}}_K^k$ be the closures of $\mathfrak{g}_K$, $\tilde{\mathfrak{g}}_K$ in $\mathfrak{g}^k$ respectively. Then, we see $\mathfrak{g} = \mathfrak{g}_K \oplus \tilde{\mathfrak{g}}_K$, $\mathfrak{g}^k = \mathfrak{g}_K^k \oplus \tilde{\mathfrak{g}}_K^k$. The strong ILB-Lie group structure of G_K is given by the restriction of ξ' to $W_1 \cap \mathfrak{g}_K$, and by definition G_K is a strong ILB-Lie subgroup of G.

3.2.4 Corollary Let K be a compact subgroup of $\mathcal{D}_{\mathcal{F}}$. Then, $\mathcal{D}_{\mathcal{F},K} = \{ \varphi \in \mathcal{D}_{\mathcal{F}} : \varphi h = h\varphi \text{ for all } h \in K \}$ is a strong ILB- and a strong ILH- Lie subgroup of $\mathcal{D}_{\mathcal{F}}$.

Proof. We use the same ILB- (resp. ILH-) coordinate $(V_F \oplus V_E \cap \gamma^1(E), \xi')$ (resp. $(V_F \oplus V_E \cap \Gamma^d(E), \xi')$ as in II.3. Use 3.2.2 and the remark there. Then, apply the above theorem, and we get the desired result.

III.3 The implicit function theorem

Let $\{\mathbb{E}, E^k, k \in N(d)\}$ be a Sobolev chain and E_1 a closed subspace of $\mathbb{E}$. Let E_1^k denotte the closure of $\mathbb{E}_1$ in E^k. Suppose there is an ILB-normal splitting $\mathbb{E} = \mathbb{E}_1 \oplus \mathbb{E}_2$ and the inequality $\| u + v \|_k \geq C\{\|u\|_k + \|v\|_k\} - D_k\{\|u\|_{k-1} + \|v\|_{k-1}\}$ in the definition of ILB-normal splittings (cf. I.4) holds for $k \geq d + 1$.

Let $\{ \mathbb{F}, F^k, k \in N(d)\}$ be another Sobolev chain with an ILB-normal splitting $\mathbb{F} = \mathbb{F}_1 \oplus \mathbb{F}_2$ and assume the inequality $\|u + v\|_k \geq C\{\|u\|_k + \|v\|_k\} + D_k\{\|u\|_{k-1} + \|v\|_{k-1}\}$ holds for every $k \geq d + 1$. Let π be the projection $\mathbb{F} \mapsto \mathbb{F}_1$ in accordance with the above splitting. Obviously, π can be extended to the projection of F^k onto F_1^k for every $k \in N(d)$ and satisfies the inequality

$$\|\pi w\|_k \leq C'\|w\|_k + D_k'\|w\|_{k-1}, \quad k \geq d + 1.$$

Suppose we have a neighborhood U of 0 in E^d and a mapping Φ of $U \cap \mathbb{E}$ into $\mathbb{F}$ with $\Phi(0) = 0$ such that $(d\Phi)_o\mathbb{E} = \mathbb{F}_1$ and the kernel of $(d\Phi)_o : \mathbb{E} \mapsto \mathbb{F}$ is $\mathbb{E}_1$.

3.3.1 Theorem <u>Notations and assumptions being as above, assume furthermore the following conditions</u> :

(a) <u>The restriction of</u> $(d\Phi)_o$, $(d\Phi)_o : \mathbb{E}_2 \mapsto \mathbb{F}_1$ <u>can be extended to an isomorphism of</u> E_2^k <u>onto</u> F_1^k <u>for every</u> $k \in N(d)$.

(b) <u>Letting</u> J <u>be the inverse of</u> $(d\Phi)_o : \mathbb{E}_2 \mapsto \mathbb{F}_1$, <u>the mapping</u> $J'\Phi : \mathbb{E} \mapsto \mathbb{E}_2 \oplus \mathbb{F}_2$ <u>is a C^∞ILBC^2-normal mapping, where</u> $J' : \mathbb{F}_1 \oplus \mathbb{F}_2 \mapsto \mathbb{E}_2 \oplus \mathbb{F}_2$ <u>is defined by</u> $J'(u,v) = (J(u), v)$.

<u>Then, there are neighborhoods</u> U_1, V_1, V_2 <u>of zeros of</u> E_1^d, F_1^d, F_2^d <u>respectively and C^∞ILBC^2-normal mappings</u> $\Psi_1 : U_1 \cap \mathbb{E}_1 \times V_1 \cap \mathbb{F}_1 \mapsto U \cap \mathbb{E}_2$ <u>and</u> $\Psi_2 : U_1 \cap \mathbb{E}_1 \times V_1 \cap \mathbb{F}_1 \times V_2 \cap \mathbb{F}_2 \mapsto \mathbb{F}_2$ <u>such that</u> $\pi\Phi(u, \Psi_1(u,v)) \equiv v$ <u>and</u> $(1-\pi)\Phi(u, \Psi_1(u,v)) + \Psi_2(u,v,w) \equiv w$.

<u>Moreover, the mapping</u> $\Psi' : U_1 \cap \mathbb{E}_1 \times V_1 \cap \mathbb{F}_1 \times V_2 \cap \mathbb{F}_2 \mapsto \mathbb{E}_1 \oplus \mathbb{E}_2 \oplus \mathbb{F}_2$ <u>defined by</u> $\Psi'(u,v,w) = (u, \Psi_1(u,v), \Psi_2(u,v,w))$ <u>is a $C^\infty ILBC^2$-normal mapping satisfying the conditions of 3.1.1.</u>

Proof. We define a mapping $\Phi' : U \cap (\mathbb{E}_1 \oplus \mathbb{E}_2) \times \mathbb{F}_2 \mapsto \mathbb{E}_1 \oplus \mathbb{E}_2 \oplus \mathbb{F}_2$ by

$\Phi'(u,v,w) = (u, J\pi\Phi(u,v), (1-\pi)\Phi(u,v) + w) = (u,0,0) + (0, J'\Phi(u,v)) + (0,0,w).$

It is easy to see that Φ' is a $C^\infty ILBC^2$-normal mapping and $(d\Phi')_0 = I$. Use the inverse function theorem and we get the desired mapping Ψ' and hence the desired result.

III.4 Applications.

(A) The purpose of this section III.4.(A) is to prove Theorem F in § 0. First of all we have to fix our terminology. We call a strong ILB-Lie group G <u>acts smoothly</u> on a finite dimensional manifold N, if the following conditions are satisfied :

(a) There exists a mapping $\rho : G \times N \mapsto N$ such that $\rho(g,\rho(h,x)) = \rho(gh,x)$ and ρ can be extended to a C^{k-d}-mapping of $G^k \times N$ into N for any $k \in N(d)$.

(b) For any fixed $x \in N$, $\rho_x(g) = \rho(g,x)$ is a smooth mapping of G^k into N for any $k \in N(d)$.

<u>Remark</u> $\mathcal{D}(M)$ acts smoothly on M.

Now, we restate the theorem.

3.4.1 Theorem <u>Let</u> G <u>be a strong ILB-Lie group acting smoothly on a finite dimensional manifold</u> N. <u>Then the isotropy subgroup</u> G_m <u>at</u> $m \in N$ <u>is a strong ILB-Lie subgroup of</u> G <u>and the orbit</u> G(m) <u>is a smooth submanifold of</u> N.

Proof. Let $\mathfrak{g}^k$ be the tangent space of G^k at e. We put $\mathfrak{g} = \cap \mathfrak{g}^k$ with the inverse limit topology. Any $u \in \mathfrak{g}^k$ induces a C^{k-d}-vector field on N, and this will be denoted by the same notation u. We put $\mathfrak{g}_0(m) = \{ u \in \mathfrak{g} : u(m) = 0 \}$, $\mathfrak{m}(m) = \{ u(m) : u \in \mathfrak{g} \}$ and $\mathfrak{n}(m)$ be a complementary subspace of $\mathfrak{m}(m)$ in the tangent space T_mN at $m \in N$. Obviously $T_mN = \mathfrak{m}(m) \oplus \mathfrak{n}(m)$ and since $\dim \mathfrak{g}/\mathfrak{g}_0(m) \leq n = \dim N < \infty$, there is $d' \in N(d)$ such that $\dim \mathfrak{g}/\mathfrak{g}_0(m) =$

$\dim \mathfrak{g}^k/\mathfrak{g}_o^k(m)$ for any $k \geq d'$, where $\mathfrak{g}_o^k(m)$ is the closure of $\mathfrak{g}_o(m)$ in $\mathfrak{g}^k$. (Remark that $\dim \mathfrak{g}^k/\mathfrak{g}_o^k(m) \geq \dim \mathfrak{g}^{k-1}/\mathfrak{g}_o^{k-1}(m)$ in general.) So, there is a finite dimensional subspace $\mathfrak{M} \subset \mathfrak{g}$ such that $\mathfrak{g} = \mathfrak{g}_o(m) \oplus \mathfrak{M}$ and $\mathfrak{g}^k = \mathfrak{g}_o^k(m) \oplus \mathfrak{M}$ for $k \geq d'$ (ILB-splitting). The space $\mathfrak{M}(m)$ can be naturally identified with $\mathfrak{M}$. We denote by $J : \mathfrak{M}(m) \mapsto \mathfrak{M}$ the identification.

Since $\mathfrak{M}(m)$, $\mathfrak{N}(m)$ are of finite dimension, we define euclidean norms on them. Now, we define another norm on $\mathfrak{g}^k$. Since $\mathfrak{g}_o^k(m) \subset \mathfrak{g}_o^{k-1}(m)$, there is a constant $C_k > 0$ such that $\|u\|_{k-1} \leq C_k\|u\|_k$ for any $u \in \mathfrak{g}_o^k(m)$. We define a new norm $\| \ \|'$ on $\mathfrak{g}_o^k(m)$ by $\|u\|'_{d'} = \|u\|_{d'}$, $\|u\|'_k = C_k \cdots C_{d'+1} \|u\|_k$. Then, we have $\|u\|'_{k+1} \geq \|u\|'_k$. This new norm $\| \ \|'_k$ gives the same topology for $\mathfrak{g}_o^k(m)$ as the original one. Define a new norm $[\![\ ]\!]_k$ on $\mathfrak{g}^k$ by $[\![u + v]\!]_k^2 = \|u\|_k'^2 + \|v\|^2$, $u \in \mathfrak{g}_o^k(m)$, $v \in \mathfrak{M}$, where $\|v\|$ is the euclidean norm on $\mathfrak{M}(m)$ identified with $\mathfrak{M}$ through J. This new norm does not change the topology on $\mathfrak{g}^k$, because $\mathfrak{M}$ is of finite dimension. We see obviously $[\![u]\!]_k \geq [\![u]\!]_{k-1} \geq [\![u]\!]_{d'}$ for any $u \in \mathfrak{g}^k$.

Let Exp be an exponential mapping defined by a smooth riemannian metric on N. We put $\Phi(u) = \mathrm{Exp}_m^{-1}\rho_m(\xi(u))$, where $\xi : U \cap \mathfrak{g} \mapsto G$ is an ILB-coordinate of G at e. For a sufficiently small U in $\mathfrak{g}^{d'}$, $\Phi(u)$ is well-defined.

Let π be the projection of $T_mN = \mathfrak{M}(m) \oplus \mathfrak{N}(m)$ onto $\mathfrak{M}(m)$.

Since $\pi(d\Phi)_o : \mathfrak{M} \mapsto \mathfrak{M}(m)$ is an isomorphism, we may assume that $\pi(d\Phi)_u : \mathfrak{M} \mapsto \mathfrak{M}(m)$ is an isomorphism for every $u \in U$, where U is a small neighborhood of 0 in $\mathfrak{g}^{d'}$.

Since $J\pi\Phi : U \cap \mathfrak{g}^{d'} \mapsto \mathfrak{M}$ is smooth, there are an open star-shaped neighborhood W' of 0 in $U \cap \mathfrak{g}^{d'}$ and a positive constant K such that $[\![J\pi(d\Phi)_u v]\!]_{d'} \leq K[\![v]\!]_{d'}$, $[\![J\pi(d^2\Phi)_u(v,w)]\!]_{d'} \leq K[\![v]\!]_{d'}[\![w]\!]_{d'}$ for any $u \in W'$. Remark that $[\![J\pi(d\Phi)_u v]\!]_k = \|J\pi(d\Phi)_u v\|$ and $[\![J\pi(d^2\Phi)_u(v,w)]\!]_k = \|J\pi(d\Phi)_u(v,w)\|$ for any $k \geq d'$. Since $[\![v]\!]_k \geq [\![v]\!]_{d'}$, it is easy to see that $J'\Phi$ is a C^∞ILBC^2-normal mapping, where $J' : \mathfrak{M}(m) \oplus \mathfrak{N}(m) \mapsto \mathfrak{M} \oplus \mathfrak{N}(m)$ is the mapping defined by $J'(u,v) = (J(u),v)$. Thus, we can use the implicit function theorem (3.3.1), where we have to change d by d'.

Consequently, there are convex neighborhoods U_1, V_1, V_2 of zeros in

$\mathfrak{g}_o^{d'}(m)$, $\mathfrak{m}(m)$, $\mathfrak{n}(m)$ respectively and $C^\infty ILBC^2$-normal mappings $\Psi_1 : U_1 \cap \mathfrak{g}_o(m) \times V_1 \longmapsto \mathfrak{m}$ and $\Psi_2 : U_1 \cap \mathfrak{g}_o(m) \times V_1 \times V_2 \longmapsto \mathfrak{n}(m)$ such that $\pi\Phi(u, \Psi_1(u,v)) \equiv v$ and $(1-\pi)\Phi(u, \Psi_1(u,v)) + \Psi_2(u,v,w) \equiv w$.

Since V_1 is naturally identified with a neighborhood of 0 of $\mathfrak{m}$, we see that $\xi(1,\Psi_1): U_1 \cap \mathfrak{g}_o(m) \times V_1 \longmapsto G$ is an ILB-coordinate of G at e and the exponential mapping $Exp_m : V_1 \times V_2 \longmapsto N$ gives a smooth local coordinate of N at m.

Let $(u,v),(u',v')$ be points in $U_1 \cap \mathfrak{g}_o(m) \times V_1$ such that $\pi\Phi(u,v) = \pi\Phi(u',v')$. Then, letting $\Phi(u,v) = (z_1,z_2)$, $\Phi(u',v') = (z_1,z_2')$, we have that $v = \Psi_1(u,z_1)$ and $v' = \Psi_1(u',z_1)$. Since U_1 is convex, $tu + (1-t)u'$ is contained in U_1 for $t \in [0,1]$ and $\pi\Phi(tu + (1-t)u', \Psi_1(tu + (1-t)u',z_1)) \equiv z_1$ for any $t \in [0,1]$.

If $z_2 \neq z_2'$, then $\Phi(tu + (1-t)u', \Psi_1(tu + (1-t)u',z_1))$ can not be constant and hence there is $s \in [0,1]$ such that

$$(d\Phi)_{(p(s),\ \Psi_1(p(s),z_1))}(u - u',\ (d\Psi_1)_{p(s)}(u - u')) \neq 0,$$

where $p(s) = su + (1-s)u'$. Therefore, we see

$$\pi(d\Phi)_{(p(s),\ \Psi_1(p(s),z_1))}(u - u',\ (d\Psi_1)_{p(s)}(u - u')) \neq 0.$$

This is because $\pi(d\Phi)_w : \mathfrak{m} \longmapsto \mathfrak{m}(m)$ is an isomorphism for every $w \in U$ and hence $(d\Phi)_w\mathfrak{g} = (d\Phi)_w\mathfrak{m}$ for any $w \in U \cap \mathfrak{g}$. (Remark that $\dim(d\Phi)_w\mathfrak{g} = \dim(d\Phi)_o\mathfrak{g} = \dim \mathfrak{m}$.)

However, the last result contradicts the fact $\Phi(p(t), \Psi_1(p(t),z_1) \equiv z_1$, hence we have $z_2 = z_2'$. Thus, we have that $\pi\Phi(u,v) = \pi\Phi(u',v')$ if and only if $\Phi(u,v) = \Phi(u',v')$. Especially, the set $\Phi = 0$ is given by $\{(u,\Psi_1(u,0)) : u \in U_1 \cap \mathfrak{g}_o(m)\}$. This implies that G_m is a strong ILB-Lie subgroup of G. More precisely, there are neighborhoods U_1', V_1' of zeros of U_1, V_1 and a $C^\infty ILBC^2$-normal mapping Ψ'' of $U_1' \cap \mathfrak{g}_o(m) \times V_1'$ into $U_1 \cap \mathfrak{g}_o(m) \times V_1$ such that $\Psi''(0,0) = (0,0)$, $(d\Psi'')_{(0,0)} = I$ and $\Psi''(u,0) = (u, \Psi_1(u,0))$. (Cf. the inverse function theorem.)

By the above argument, we see also that $\Phi(u,v)$ is determined by $\pi\Phi(u,v)$. Thus, the image $\Phi(U_1 \cap \mathfrak{g}_o(m) \times V_1)$ is equal with $\Phi(0, V_1)$, because $\pi\Phi : V_1 \longmapsto \mathfrak{m}(m)$ is an into-diffeomorphism. Moreover, letting $z_1 = \pi\Phi(0,v)$, we see $\Phi(0,v) = (z_1, -\Psi_2(0,z_1,0))$. Thus, the image $\Phi(U_1 \cap \mathcal{G}_0(m) \times V_1)$ is given by

$$\{ (z_1, -\Psi_2(0,z_1,0)) : z_1 \in V_1 \}.$$

This implies the orbit $G(m)$ is a smooth submanifold of N.

Remark $\mathcal{D}(M,S)$ in II.4 acts smoothly on M.

(B) Now, we back to the setting of II.3. $\mathcal{F}$ is a smooth fibering of M with a compact fibre and the projection $\pi : M \mapsto N$. F is the subbundle of T_M defined by the tangent spaces of the fibres of $\mathcal{F}$ and E a complementary subbundle. $\mathcal{D}_{\mathcal{F}}$, $\Gamma_{\mathcal{F}}(E)$ are as in II,3. Define the projection $\tilde{\pi} : \mathcal{D}_{\mathcal{F}} \longmapsto \mathcal{D}(N)$ by $\pi\varphi(x) = \tilde{\pi}(\varphi)\pi(x)$ for $\varphi \in \mathcal{D}_{\mathcal{F}}$.

We use here a riemannian metric such that F and E are perpendicular to each other. Then, $< \nabla^s u, \nabla^s v > = 0$ for any $u \in \Gamma(F)$, $v \in \Gamma(E)$. Thus, we can see that the splitting $\Gamma(T_M) = \Gamma(F) \oplus \Gamma(E)$ is ILB-normal and also ILH-normal.

Let $(V_F \oplus V_E , \xi')$ be the same ILB- (resp. ILH-) coordinate of $\mathcal{D}_{\mathcal{F}}$ at e as in 2.3.1 Lemma. In the defintion of ξ', we use an exponential mapping Exp' defined on N. Using this Exp', let (U', ξ) be the ILB- (resp. ILH-)coordinate of $\mathcal{D}(N)$ at e such that $\xi(u)(x) = \text{Exp}'u(x)$.

Now, the local expression of $\tilde{\pi}$ is given by $\Phi(u,v) = \xi^{-1}\tilde{\pi}\xi'(u + v)$, where Φ is regarded as a mapping of $V_F \cap \Gamma(F) \oplus V_E \cap \Gamma_{\mathcal{F}}(E)$ into $U' \cap \Gamma(T_N)$. Then, $\Phi(u,v) = d\pi\, v$ by virtue of the specific choice of ξ', ξ. It is now easy to see that Φ satisfies all of the conditions of the implicit function theorem 3.3.1.

3.4.2 Theorem $\mathcal{D}_{[\mathcal{F}]} = \tilde{\pi}^{-1}(e)$ <u>is a strong ILB- and a strong ILH- Lie subgroup of</u> $\mathcal{D}_{\mathcal{F}}$. <u>Moreover, for any strong</u> ILB- (resp, <u>strong ILH-) Lie subgroup</u> G <u>of</u> $\mathcal{D}(N)$, $\tilde{\pi}^{-1}(G)$ <u>is a strong ILB- (resp. a strong ILH-) Lie subgroup of</u> $\mathcal{D}_{\mathcal{F}}$.

§ IV Vector bundle over strong ILB-Lie groups

IV.1 Definition of vector bundle over strong ILB-Lie groups.

In [30], the author defined vector bundles over groups of diffeomorphisms and using these, he obtained the main theorem of [31]. Here, we will discuss about an abstract treatment of vector bundles over strong ILB-Lie groups and will discuss about semi-direct products.

Let G be a strong ILB-Lie group with the Lie algebra $\mathfrak{g}$ and let $\mathfrak{g}^k$ be the tangent space of G^k at e. The system $\{\mathfrak{g}, \mathfrak{g}^k, k \in N(d)\}$ is then a Sobolev chain. Let $\{\mathbb{F}, \mathbb{F}^k, k \in N(d)\}$ be another Sobolev chain. We consider the following mapping $\tilde{T} : \mathbb{F} \times \tilde{U} \cap G \times \tilde{U} \cap G \mapsto \mathbb{F}$ for an open neighborhood $\tilde{U}$ of e in G^d :

(VB,1) $\tilde{T}$ is linear with respect to the first variable and putting $\hat{T}(g,h) = \tilde{T}(*,g,h)$, $\hat{T}$ satisfies $T(g,e) = \mathrm{id}$, and $\hat{T}(gh,h')\hat{T}(g,h) = \hat{T}(g,hh')$ whenever they are defined.

(VB,2) $\tilde{T}$ can be extended to a C^{ℓ}-mapping of $F^{k+\ell} \times \tilde{U} \cap G^{k+\ell} \times \tilde{U} \cap G^k$ into F^k for every $k \in N(d)$, $\ell \geq 0$.

(VB,3) If the third variable h is fixed in $\tilde{U} \cap G^k$, then the extended mapping $\tilde{T} : F^k \times \tilde{U} \cap G^k \times \{h\} \mapsto F^k$ is of class C^{∞}.

<u>Remark</u> If $\hat{T}(g,h)$ is independent from g, then $\hat{T}(hh') = \hat{T}(h')\hat{T}(h)$. Thus, $\hat{T}$ gives a representation of the local group $\tilde{U} \cap G$.

The following lemma gives an example of such a mapping :

4.1.1 Lemma <u>Notations being as in (N,5) in § I, $\tilde{T}_\theta(w,g,h) = \theta(w,\xi^{-1}(g),\xi^{-1}(h))$ satisfies the above conditions</u> (VB,1 ~3).

Proof. Since $\theta(w,u,v) = (d\eta_v)_u w$, we have $(d\eta_o)_u = \mathrm{id.}$ and that $(d\eta_{v'})_{\eta(u,v)}(d\eta_v)_u = (d\eta_{\eta(v,v')})_u$. Thus, $\tilde{T}_\theta$ satisfies (VB,1). The condition (VB,2) is the same as

(N,5). If v is fixed, then η_v is smooth (cf. (N,4)) and hence so is $d\eta_v$. This implies the condition (VB,3).

Since G^d is a topological group, there is an open neighborhood $\widetilde{W}$ of e in G^d such that $\widetilde{W}^{-1} = \widetilde{W}$ and $\widetilde{W}^2 \subset \widetilde{U}$. By the equality $\hat{T}(gh,h^{-1})\hat{T}(g,h) = \hat{T}(g,h) = \mathrm{id}$, we see that $\hat{T}(g,h) : F^k \mapsto F^k$ is invertible for every $g, h \in \widetilde{W} \cap G^k$.

For two mappings $\widetilde{T} : \mathbb{F} \times \widetilde{U} \cap G \times \widetilde{U} \cap G \mapsto \mathbb{F}$, $\widetilde{T}' : \mathbb{F} \times \widetilde{U}' \cap G \times \widetilde{U}' \cap G \mapsto \mathbb{F}$ satisfying (VB,1 - 3), these are said to be <u>equivalent</u>, if there is an open neighborhood $\widetilde{V}$ of e in $\widetilde{U} \cap \widetilde{U}'$ such that $\mu(x)w = \hat{T}'(e,x)\hat{T}(e,x)^{-1}w$ can be extended to a C^∞-mapping of $\widetilde{V} \cap G^k \times F^k$ onto F^k and $\mu(x) : F^k \mapsto F^k$ is an isomorphism for every $x \in \widetilde{V} \cap G^k$. We use the notation $\widetilde{T} \sim \widetilde{T}'$ or $\hat{T} \sim \hat{T}'$.

Now, the vector bundle $B(\mathbb{F},G,\widetilde{T})$, that we want to define here by using the equivalence class of $\widetilde{T}$ is topologically the direct product, that is, $B(\mathbb{F},G,\widetilde{T})$ is homeomorphic to $\mathbb{F} \times G$. However, we use a different local trivialization and make a system $B(F^{k'},G^k,\widetilde{T})$ for any $k \geq k' \geq d$. In fact, we take the following local trivialization :

$$\tau_g : (\widetilde{W} \cap G)g \times \mathbb{F} \mapsto \mathbb{F} \times G, \qquad \tau_g(xg)w = (\hat{T}(e,x)^{-1}w,\ xg).$$

If $xg = yh$, then the transition function $t_{h,g}(xg) = \tau_h(yh)^{-1}\tau_g(xg)$ is given by

$$t_{h,g}(xg) = \hat{T}(e,y)\hat{T}(e,x)^{-1} = \hat{T}(xx^{-1},y)\hat{T}(x,x^{-1}) = \hat{T}(x,x^{-1}y) = \hat{T}(x,gh^{-1}).$$

Thus, by the property (VB,3), we see that $t_{h,g} : (\widetilde{W} \cap G)g \times \mathbb{F} \mapsto \mathbb{F}$ can be extended to the smooth mapping of $(\widetilde{W} \cap G^k)g \times F^{k'}$ onto $F^{k'}$ for every k, k' such that $k \geq k' \geq d$. So, by this transition function $t_{h,g} : (\widetilde{W} \cap G^k)g \times F^{k'} \mapsto F^{k'}$, we define a system of smooth vector bundles $B(F^{k'},G^k,\widetilde{T})$ over G^k with the fibre $F^{k'}$. $B(F^{k'},G^k,\widetilde{T})$ is then a smooth Banach vector bundle over G^k, and the pull back of the bundle $B(F^{k'},G^{k'},\widetilde{T})$ by the inclusion $G^k \subset G^{k'}$. We put $B(\mathbb{F},G,\widetilde{T}) = \cap B(F^k,G^k,\widetilde{T})$ with the inverse limit topology, and call it an <u>ILB-vector bundle over</u> G defined by $\widetilde{T}$. If all model spaces F^k are Hilbert spaces, then we call it an ILH-vector bundle over G.

τ_g is called a local ILB- (or ILH-) trivialization of $B(\mathbb{F},G,\tilde{T})$. This can be obviously extended to a local trivialization of $B(F^{k'},G^k,\tilde{T})$.

It is clear that $B(\mathbb{F},G,\tilde{T})$ depends only on the equivalence class of $\tilde{T}$, that is, if $\tilde{T} \sim \tilde{T}'$, then the identity mapping $\iota : \mathbb{F} \times G \mapsto \mathbb{F} \times G$ can be extended to an isomorphism of $B(F^{k'},G^k,\tilde{T})$ onto $B(F^{k'},G^k,\tilde{T}')$ for every $k \geq k' \geq d$.

The tangent bundle T_G of a strong ILB-Lie group G is an ILB-vector bundle over G defined by $\tilde{T}_\theta$ in 4.1.1. If $\tilde{T}(w,g,h)$ does not depend on g, then $B(F^k,G^k,\tilde{T})$ is a trivial vector bundle and the trivialization does not depend on k.

Let $\tilde{R} : \mathbb{F} \times G \times G \mapsto \mathbb{F} \times G$ be the mapping defined by $\tilde{R}(w,g,g') = (w,gg')$. Then, we have

4.1.2 Lemma $\tilde{R}$ can be extended to the C^ℓ-mapping of $B(F^{k'+\ell},G^{k+\ell},\tilde{T}) \times G^k$ into $B(F^{k'},G^k,\tilde{T})$ for every $k, k' \in N(d)$ such that $k \geq k'$. If the third variable g' is fixed, then the mapping $\tilde{R}_{g'}$ defined by $\tilde{R}_{g'}(w,g) = \tilde{R}(w,g,g')$ is a smooth mapping of $B(F^{k'},G^k,\tilde{T})$ onto itself for every $k \geq k' \geq d$, and for every $g' \in G^k$.

Proof. Let $\tilde{W}_1$ be an open neighborhood of e of G^d. We have only to show the smoothness property on open subsets $(\tilde{W}_1 \cap G^{k+\ell})g$, $(\tilde{W}_1 \cap G^k)h$ in $G^{k+\ell}$, G^k respectively, where $g,h \in G$. Let $\tilde{W}$ be as above. For a small $\tilde{W}_1$, we may assume that $g\tilde{W}_1g^{-1} \subset \tilde{W}$. Using the local trivializations τ_g, τ_h, τ_{gh}, the local expression of $\tilde{R}$ is given by $\tau_{gh}(xgyh)^{-1}(\tau_g(xg)w,xgyh)$. This is equal to

$$(\hat{T}(e,xgyg^{-1})\hat{T}(e,x)^{-1}w,\ xgyg^{-1}gh) = (\tilde{T}(w,x,gyg^{-1}),xgyg^{-1}gh).$$

Thus, by the property (VB,2), we have the first half. If y is fixed, then the property (VB,3) yields the second half.

IV.2 An example of $\tilde{T}$.

Here, we will give a generalization of 4.1.1 Lemma. Let F be a smooth finite dimensional riemannian vector bundle over a closed manifold M. We define a riemannian connection on F and denote by $\tau(\operatorname{Exp} X)$ the parallel displacement along the curve $\operatorname{Exp} tX$, $t \in [0,1]$.

Now, first of all we remark that although all argument in this section will be given by using H-norms, the same results hold for B-norms changing $\Gamma^k(F)$ by $\gamma^k(F)$ and $d = \dim M + 5$ by $d = 1$.

Let $\xi : U \cap \Gamma(T_M) \mapsto \mathcal{D}$ be an ILH-coordinate mapping defined by $\xi(u)(x) = \operatorname{Exp} u(x)$ as in § II. Let $\tau(\operatorname{Exp} u(\operatorname{Exp} v(x))$ be the parallel displacement along the curve $\operatorname{Exp} tu(\operatorname{Exp} v(x))$, $t \in [0,1]$. Let V be the same neighborhood of 0 in $\Gamma^d(T_M)$ as in (N,2) in § I.

4.2.1 Lemma <u>Put</u> $T_F(w,u,v)(x) = \tau(\operatorname{Exp}\eta(u,v)(x))^{-1}\,\tau(\operatorname{Exp} u(\operatorname{Exp} v(x)))w(\operatorname{Exp} v(x))$ <u>for every</u> $u,v \in V \cap \Gamma(T_M)$. <u>Then,</u> $\tilde{T}_F(w,g,h) = T_F(w,\xi^{-1}g,\xi^{-1}h)$ <u>satisfies the conditions</u> (VB,1 - 3).

Proof.

$$(\hat{T}_F(\eta(u,v),v')\hat{T}_F(u,v)w)(x)$$

$$= \tau(\operatorname{Exp}\eta(\eta(u,v),v')(x)))^{-1}\tau(\operatorname{Exp} u(\operatorname{Exp} v(\operatorname{Exp} v'(x)))w(\operatorname{Exp} v(\operatorname{Exp} v'(x)))$$

$$= \tau(\operatorname{Exp}\eta(u,\eta(v,v'))(x)^{-1}\tau(\operatorname{Exp} u(\operatorname{Exp}\eta(v,v')(x)))w(\operatorname{Exp}\eta(v,v')(x))$$

$$= (\hat{T}_F(u,\eta(v,v'))w)(x).$$

Thus, $\tilde{T}_F$ has the property (VB,1). To prove the properties (VB,2 - 3), we put

$$T_F(w,u,v)(x) = \tau(\operatorname{Exp}\eta(u,v)(x))^{-1}\tau(\operatorname{Exp} u(\operatorname{Exp} v(x)))\tau(\operatorname{Exp} v(x))\tau(\operatorname{Exp} v(x))^{-1}w(\operatorname{Exp} v(x)).$$

By 2.1.2, we see $\tau(\operatorname{Exp} v(x))^{-1}w(\operatorname{Exp} v(x)) = R'(w,v)(x)$ and hence this can be extended to the C^ℓ-mapping of $\Gamma^{k+\ell}(F) \times V \cap \Gamma^k(T_M)$ into $\Gamma^k(F)$ for any $k \in N(d)$, $d = \dim M + 5$. Thus, we have only to show the mapping

$$T_\Delta(w,u,v) = \tau(\operatorname{Exp}\eta(u,v)(x))^{-1}\tau(\operatorname{Exp} u(\operatorname{Exp} v(x)))\tau(\operatorname{Exp} v(x))w(x)$$

can be extended to the C^ℓ-mapping of $\Gamma^{k+\ell}(F) \times V \cap \Gamma^{k+\ell}(T_M) \times V \cap \Gamma^k(T_M)$ into $\Gamma^k(F)$. Now, let $\Xi(u,v)(x) = \operatorname{Exp}_x^{-1}(\operatorname{Exp}\tau(\operatorname{Exp} v(x))u(x))$, where τ is the parallel displacement in T_M along the curve $\operatorname{Exp} tv(x)$, $t \in [0,1]$, and let

$$\tilde{\Xi}(w,u,v)(x) = \tau(\operatorname{Exp}\Xi(u,v)(x))^{-1}\tau(\operatorname{Exp}\tau(\operatorname{Exp} v(x))u(x))\tau(\operatorname{Exp} v(x))w(x).$$

Then, Ξ and $\tilde{\Xi}$ are defined from smooth fibre preserving mappings, and therefore they can be extended to smooth mappings of $V \cap \Gamma^k(T_M) \times V \cap \Gamma^k(T_M)$ into $\Gamma^k(T_M)$ and of $\Gamma^k(F) \times V \cap \Gamma^k(T_M) \times V \cap \Gamma^k(T_M)$ into $\Gamma^k(T_M)$ respectively (cf. 2.1.3). Remark that $\eta(u,v)(x) = \mathrm{Exp}_x^{-1}(\mathrm{Exp}u(\mathrm{Exp}v(x)) = \Xi(R'(u,v),v)$. Therefore, we have $T_\Delta(w,u,v) = \tilde{\Xi}(w,R'(u,v),v)$. Thus, by 2.1.2, we have the desired result.

Remark Let $B(\Gamma(F), \mathfrak{D}, \tilde{T}_F)$ be the ILH-vector bundle over the strong ILH-Lie group $\mathfrak{D}$ defined by $\tilde{T}_F$ above. Then, this is the same bundle which was denoted by $\gamma(F)$ in the previous papers [30,31].

IV.3 Invariant bundle morphisms.

Here, we will give the notion of ILB-subbundle of $B(\mathbb{F},G,\tilde{T})$ and a sufficient condition to get such ILB-subbundles.

First of all, we have to remark that $\hat{T}(g,h) = \tau_e(gh)^{-1}\tilde{R}_h\tau_e(g) = \hat{T}(e,gh)\hat{T}(e,g)^{-1}$. Thus, $\hat{T}'(g,h) = \mu(gh)\hat{T}(g,h)\mu(g)^{-1}$ is equivalent with $\hat{T}$, if μ satisfies that $\mu(e) =$ id. and that $\mu : \tilde{V} \cap G \times \mathbb{F} \mapsto \mathbb{F}$ can be extended to a smooth mapping of $\tilde{V} \cap G \times \mathbb{F}^k$ onto $\mathbb{F}^k$, $k \in N(d)$, such that $\mu(g) : \mathbb{F}^k \mapsto \mathbb{F}^k$ is an isomorphism for every $g \in \tilde{V} \cap G^k$.

An ILB-vector bundle $B(\mathbb{F}_1,G,\tilde{T}_1)$ is called an ILB-subbundle of $B(\mathbb{F},G,\tilde{T})$, if the following conditions are satisfied :

(SB,1) $\mathbb{F}_1$ is a closed subspace of $\mathbb{F}$ and there is an ILB-splitting $\mathbb{F} = \mathbb{F}_1 \oplus \mathbb{F}_2$.

(SB,2) There is $\tilde{T}'$ such that $\tilde{T}' \sim \tilde{T}$ and $\tilde{T}'|\mathbb{F}_1 \equiv \tilde{T}_1$, that is, $\tilde{T}'(w,g,h) \equiv \tilde{T}_1(w,g,h)$ for $w \in \mathbb{F}_1$.

Let $B(\mathbb{E},G,\tilde{T})$, $B(\mathbb{F},G,\tilde{T}')$ be ILB-vector bundles over G. By 4.1.2 Lemma, the group G operates from the right hand side on $B(\mathbb{E},G,\tilde{T})$, $B(\mathbb{F},G,\tilde{T}')$ respectively. So, if $\tilde{A} : B(\mathbb{E},G,\tilde{T}) \mapsto B(\mathbb{F},G,\tilde{T}')$ is a right invariant bundle morphism, then $\tilde{A}$ induces a linear mapping $A : \mathbb{E} \mapsto \mathbb{F}$ by identifying $\mathbb{E}$, $\mathbb{F}$ with the fibres at the identity. Conversely, starting with a linear mapping $A : \mathbb{E} \mapsto \mathbb{F}$, we can make a right invariant fibre preserving mapping $\tilde{A} : B(\mathbb{E},G,\tilde{T}) \mapsto B(\mathbb{F},G,\tilde{T}')$.

In this section, we consider a right invariant fibre preserving mapping $\tilde{A}$ which satisfies the following conditions :

(a) Let r be a non-negative integer. $\tilde{A} : B(\mathbb{E},G,\tilde{T}) \longrightarrow B(\mathbb{F},G,\tilde{T}')$ can be extended to a C^∞-bundle morphism of $B(E^{k'},G^k,\tilde{T})$ into $B(F^{k'-r},G^k,\tilde{T}')$ for every $k \geq k' \geq d+r$.

(b) Let $\mathrm{Ker}A = \mathbb{E}_1$ and $\mathrm{Im}A = \mathbb{F}_1$. There are ILB-splittings $\mathbb{E} = \mathbb{E}_1 \oplus \mathbb{E}_2$ and $\mathbb{F} = \mathbb{F}_1 \oplus \mathbb{F}_2$, and $AE^{k+r} = F_1^k$ for every $k \in N(d)$.

(c) Let $p : \mathbb{F} \mapsto \mathbb{F}_1$ be the projection in accordance with the above ILB-splitting. Then, $\|pv\|_k \leq C\|v\|_k + D_k\|v\|_{k-1}$, $k \in N(d+r)$, where C, D_k are positive constants and C does not depend on k. (This implies that the splitting $\mathbb{F} = \mathbb{F}_1 \oplus \mathbb{F}_2$ is an ILB-normal splitting.)

(d) $\|Av\|_{k-r} \geq C'\|v\|_k - D'_k\|v\|_{k-1}$, $k \in N(d+r)$, for every $v \in \mathbb{E}_2$, where C', D'_k are positive constants and C' does not depend on k.

(e) Let (U,ξ) be an ILB-coordinate of G at e such that the mapping $\tilde{T}$, $\tilde{T}'$ are defined on $\xi(U) \cap G$. Putting $\Psi(u)w = \hat{T}'(e,\xi(u))A\hat{T}(e,\xi(u))^{-1}w$ (that is, $= \tau'_e(\xi(u))^{-1}\tilde{A}\tau_e(\xi(u))w$, a local expression of $\tilde{A}$), there is a neighborhood W of 0 in $\mathfrak{g}^{d+r}$ such that Ψ satisfies the following inequalities for any $u \in W \cap \mathfrak{g}$ and $w \in \mathbb{E}$:

$$\|\Psi(u)w\|_{k-r} \leq C\{\|u\|_k\|w\|_d + \|w\|_k\} + P_k(\|u\|_{k-1})\|w\|_{k-1},$$

$$\|(d_1\Psi)_u(v,w)\|_{k-r} \leq C\{\|u\|_k\|v\|_d\|w\|_d + \|v\|_k\|w\|_d + \|v\|_d\|w\|_k\} + P_k(\|u\|_{k-1})\|v\|_{k-1}\|w\|_{k-1},$$

$$\begin{aligned}\|(d_1^2\Psi)_u(v_1,v_2,w)\|_{k-r} \leq C\{&\|u\|_k\|v_1\|_d\|v_2\|_d\|w\|_d + \|v_1\|_k\|v_2\|_d\|w\|_d + \|v_1\|_d\|v_2\|_k\|w\|_d \\ &+ \|v_1\|_d\|v_2\|_d\|w\|_k\} \\ &+ P_k(\|u\|_{k-1})\|v_1\|_{k-1}\|v_2\|_{k-1}\|w\|_{k-1}, \quad k \in N(d+r),\ k \geq d+1,\end{aligned}$$

where C is a positive constant independent from k and P_k is a polynomial with positive coefficients, and $d_1\Psi$, $d_1^2\Psi$ mean the first and the second derivatives of Ψ with respect to the first variable.

The last condition (e) is fairly complicated. However, this is similar to that of C^∞ILBC^2-normal mappings. So, we call henceforth that $\tilde{A}$ is a <u>C^∞ILBC^2-normal bundle morphism of order</u> r, if $\tilde{A}$ satisfies (a) and (e).

Now, start with such a bundle morphism $\widetilde{A}$. The goal here is to prove the following :

4.3.1 Theorem <u>Let $\widetilde{A} : B(\mathbb{E},G,\widetilde{T}) \mapsto B(\mathbb{F},G,\widetilde{T}')$ be a right invariant C^∞ILBC^2-normal bundle morphism of order r satisfying (b) ~ (d) above. Then, Ker$\widetilde{A}$ is an ILB-subbundle of $B(\mathbb{E},G,\widetilde{T})$ by restricting the index k in N(d+r) and Im$\widetilde{A}$ is an ILB-subbundle of $B(\mathbb{F},G,\widetilde{T}')$ by sliding the indeces of the fibre F^k from k to k-r.</u> (Cf. <u>the remark below.</u>) <u>Moreover, if we restrict the range and regard $\widetilde{A}$ as a bundle morphism of $B(\mathbb{E},G,\widetilde{T})$ onto the subbundle Im$\widetilde{A} = B(\mathbb{F}_1,G,\widetilde{T}^*_1)$, then this is also a C^∞-ILBC^2-normal bundle morphism.</u>

<u>Remark</u> If we use the Sobolev chain $\{\mathbb{F},F^{k-r},k \in N(d+r)\}$ instead of $\{\mathbb{F},F^k,k \in N(d)\}$, then we get a system of vector bundles $B(F^{k'-r},G^k,\widetilde{T}')$, $k \geq k' \geq d+r$, with the same inverse limit $B(\mathbb{F},G,\widetilde{T}')$. Im$\widetilde{A}$ is a subbundle of the system $\{B(F^{k-r},G^k,\widetilde{T}'), k \in N(d+r)\}$ but not of $\{B(F^k,G^k,\widetilde{T}'), k \in N(d+r)\}$.

The above theorem will be proved in the several lemmas below.

Since $A = \Psi(0) = p\Psi(0)$, $p\Psi(0) : \mathbb{E}_2 \mapsto \mathbb{F}_1$ can be extended to the isomorphism of E_2^k onto F_1^{k-r} for every $k \in N(d+r)$. Let $G(0) : \mathbb{F}_1 \mapsto \mathbb{E}_2$ be its inverse. Then, G(0) can be extended to the isomorphism of F_1^{k-r} onto E_2^k. Moreover, by the inequality (d), we have

$$(G) \quad \|G(0)v\|_k \leq C''\|v\|_{k-r} + D_k''\|v\|_{k-r-1}, \quad k \in N(d+r+1),$$

and using this, we have also

$$\|G(0)p(d_1\Psi)_u(v,w)\|_k \leq C\{\|u\|_k\|v\|_d\|w\|_d + \|v\|_k\|w\|_d + \|v\|_d\|w\|_k\} + P_k(\|u\|_{k-1})\|v\|_{k-1}\|w\|_{k-1}, \qquad k \in N(d+r+1).$$

Therefore,

$$\|G(0)p\Psi(u)w - w\|_k \leq \int_0^1 \|G(0)p(d_1\Psi)_{tu}(u,w)\|_k dt$$

$$\leq \frac{C}{2}\|u\|_k\|u\|_d\|w\|_d + C\{\|u\|_d\|w\|_k + \|u\|_k\|w\|_d\} + P_k(\|u\|_{k-1})\|u\|_{k-1}\|w\|_{k-1}.$$

Thus, by the completely same method as in 3.1.2 Lemma, we see the following :

4.3.2 Lemma <u>For a sufficiently small convex neighborhood</u> W_1 <u>of</u> 0 <u>in</u> $\mathfrak{g}^{d+r}$ <u>such that</u> $W_1 \subset W$ (cf. the condition (e)), $p\Psi(u) : E_2^k \mapsto F_1^{k-r}$ <u>is an isomorphism for every</u> $u \in W_1 \cap \mathfrak{g}^k$, $k \in N(d+r)$.

Let $G(u)$ be the inverse of $p\Psi(u) : \mathbb{E}_2 \mapsto \mathbb{F}_1$ and $G(u)$ can be extended to the isomorphism of F_1^{k-r} onto E_2^k. Moreover, $G : W_1 \cap \mathfrak{g} \times \mathbb{F}_1 \mapsto \mathbb{E}_2$ can be extended to the C^∞-mapping of $W_1 \cap \mathfrak{g}^k \times F_1^{k-r}$ onto E_2^k for any $k \in N(d+r)$.

4.3.3 Lemma <u>There is an open neighborhood</u> W_2 <u>of</u> 0 <u>in</u> $\mathfrak{g}^{d+r}$ <u>such that</u> $W_2 \subset W_1$ <u>and</u> $\Psi(u)E_2^{d+r} = \Psi(u)E^{d+r}$ <u>for any</u> $u \in W_2$. <u>Moreover,</u> $\Psi(u)E_2^k = \Psi(u)E^k$ <u>for any</u> $u \in W_2 \cap \mathfrak{g}^k$, $k \in N(d+r)$.

Proof. Assume there were a sequence $\{x_n\}$ converging to 0 in $\mathfrak{g}^{d+r}$ such that $\Psi(x_n)E_2^{d+r} \subsetneqq \Psi(x_n)E^{d+r}$. Then, there exists w_n for each n such that $\Psi(x_n)w_n \in F_2^d$ and $\|\Psi(x_n)w_n\|_d = 1$. Since $\Psi(x_n)\tau_e(x_n)^{-1}\tilde{R}_{\xi(x_n)}E_1^{d+r} = \{0\}$, we may assume that $w_n \in \tau_e(x_n)^{-1}\tilde{R}_{\xi(x_n)}E_2^{d+r} = \hat{T}(e,\xi(x_n))E_2^{d+r}$.

Let $y_n = \iota(x_n) = \xi^{-1}(\xi(x_n)^{-1})$. $\{y_n\}$ converges to 0 in $\mathfrak{g}^{d+r}$. (Cf. (N,6).) Since the mapping $\tilde{T} : E^{d+r} \times \tilde{W}_1 \times \tilde{W}_1 \mapsto E^{d+r}$ is continuous and $\tilde{T}(0,e,e) = 0$ (where $\tilde{W}_1 = \xi(W_1)$), there is an open neighborhood W' of 0 in W_1 such that

$$\|\tilde{T}(w,\xi(u),\xi(v)\|_{d+r} \leq K\|w\|_{d+r}$$

for any $u, v \in W'$, where K is a positive constant. Since $\hat{T}(\xi(x_n),\xi(y_n))\hat{T}(e,\xi(x_n)) = \text{id.}$, we have

$$K^{-1}\|w\|_{d+r} \leq \|\hat{T}(e,\xi(x_n))w\|_{d+r} \leq K\|w\|_{d+r},$$

for sufficiently large n, and similarly we have

$$K^{-1}\|w\|_d \leq \|\hat{T}'(e,\xi(x_n))w\|_d \leq K\|w\|_d,$$

for sufficiently large n. Since $\Psi(x_n)w_n = \hat{T}'(e,\xi(x_n))AT(e,\xi(x_n))^{-1}w_n$, we see that there is a constant C such that $\|\Psi(x_n)w_n\|_d \geq C\|w_n\|_{d+r}$, $C > 0$. Namely $\{w_n\}$ is bounded in $\mathfrak{g}^{d+r}$. Notice that $\lim_{n\to\infty}\Psi(x_n) = \Psi(0) = A$ in the operator norm of

$L(E^{d+r},F^d)$. Putting $w_n = w_{1,n} + w_{2,n}$, $w_{i,n} \in E_i^{d+r}$, we have $\{w_{1,n}\}$ and $\{w_{2,n}\}$ are also bounded in $\mathfrak{J}^{d+r}$. Therefore,

$$\begin{aligned} 1 &= \lim \|\Psi(x_n)w_n\|_d \leq \lim \|\Psi(x_n)w_{2,n}\|_d + \lim \|(\Psi(x_n) - \Psi(0))w_{1,n}\|_d \\ &= \lim \|\Psi(x_n)w_{2,n}\|_d \leq \lim \|\Psi(0)w_{2,n}\|_d + \lim \|(\Psi(x_n) - \Psi(0))w_{2,n}\|_d \\ &= \lim \|\Psi(0)w_{2,n}\|_d \leq \lim \|p\Psi(x_n)w_{2,n}\|_d + \lim \|p(\Psi(x_n) - \Psi(0))w_{2,n}\|_d \\ &= \lim \|p\Psi(x_n)w_{2,n}\|_d \leq \lim \|p\Psi(x_n)w_n\|_d + \lim \|p(\Psi(x_n) - \Psi(0))w_{1,n}\|_d \\ &= \lim \|p\Psi(x_n)w_n\|_d = 0. \end{aligned}$$

This is a contradiction, hence there exists a neighborhood W_2 such that $\Psi(u)E_2^{d+r} = \Psi(u)E^{d+r}$ for any $u \in W_2$.

Now, let $u \in W_2 \cap \mathfrak{J}^k$. For any $w' \in \Psi(u)E^k$, there is $w \in E_2^{d+r}$ such that $\Psi(u)w = w'$. Apply the projection p. Since $w' \in F^{k-r}$, we see $pw' \in F_1^{k-r}$ and $p\Psi(u)w = pw'$. Since $p\Psi(u) : E_2^k \longmapsto F_1^{k-r}$ is an isomorphism (4.3.2), we have $w \in E_2^k$. Thus, $\Psi(u)E_2^k = \Psi(u)E^k$.

4.3.4 Corollary <u>Notations being as above, we have</u> $\mathrm{Ker}\{p\Psi(u) : E^k \longmapsto F_1^{k-r}\} = \mathrm{Ker}\{\Psi(u) : E^k \longmapsto F^{k-1}\}$ <u>for any</u> $u \in W_2 \cap \mathfrak{J}^k$, $k \in N(d+r)$.

Proof. We have only to show for the case $k = d+r$. If the above statement is false, then there is a sequence $\{x_n\}$ converging to 0 such that $\Psi(x_n)E^{d+r} \cap F_2^d \neq \{0\}$. Thus, by the same argument as above, we have the desired result.

Proof of 4.3.1 Theorem.

Define a mapping $\Phi_E : W_2 \cap \mathfrak{J} \times \mathbb{E} \longmapsto \mathbb{E}$ by $\Phi_E(u)(w_1 + w_2) = w_1 + w_2 - G(u)p\Psi(u)w_1$, $w_1 \in \mathbb{E}_1$, $w_2 \in \mathbb{E}_2$. Then, Φ_E can be extended to a C^∞-mapping of $W_2 \cap \mathfrak{J}^k \times E^k$ onto E^k and for any $u \in W_2 \cap \mathfrak{J}^k$, $\Phi_E(u)$ is an isomorphism of E^k onto itself, where $k \in N(d+r)$. Let $\hat{T}^*(\xi(u),\xi(v)) = \Phi_E(\eta(u,v))^{-1}\hat{T}(\xi(u),\xi(v))\Phi_E(u)$. Then, $\hat{T}^*$ is equivalent with $\hat{T}$. Furthermore, by 4.3.4, we see that $\Phi_E(u)\mathbb{E}_1 = \mathrm{Ker}\Psi(u) = \tau_e(\xi(u))^{-1}\tilde{R}_{\xi(u)}\mathbb{E}_1 = \hat{T}(e,\xi(u))\mathbb{E}_1$. Thus, $\hat{T}^*(\xi(u),\xi(v))\mathbb{E}_1$ is equal to

$$\Phi_E(\eta(u,v))^{-1}\hat{T}(e,\xi(u)\xi(v))\hat{T}(e,\xi(u))^{-1}\hat{T}(e,\xi(u))\mathbb{E}_1$$

hence this is equal to $\mathbb{E}_1$. Therefore, restricting $\hat{T}^*$ onto $\mathbb{E}_1$, we have the mapping $\hat{T}^*_1$ satisfying (VB, $1 \sim 3$). Thus, $\mathrm{Ker}\tilde{A}$ is an ILB-subbundle of $B(\mathbb{E}, G, \tilde{T})$.

Let $I(u) : \mathbb{F}_1 \mapsto \mathbb{F}_1$ be the inverse of $p\Psi(u)G(0) : \mathbb{F}_1 \mapsto \mathbb{F}_1$. We define a mapping $\Phi_F : W_2 \cap \mathfrak{g} \times \mathbb{F} \mapsto \mathbb{F}$ by

$$\Phi_F(u)(w_1 + w_2) = w_1 + w_2 + (1-p)\Psi(u)G(0)I(u)w_1 , \quad w_1 \in \mathbb{F}_1, \quad w_2 \in \mathbb{F}_2 .$$

Φ_F can be extended to a C^∞-mapping of $W_2 \cap \mathfrak{g}^k \times F^{k-r}$ onto F^{k-r}, $k \in N(d+r)$, and $\Phi_F(u)$ is an isomorphism of F^{k-r} onto itself for any $u \in W_2 \cap \mathfrak{g}^k$. Furthermore, by 4.3.3 Lemma, we have $\Phi_F(u)\mathbb{F}_1 = \Psi(u)\mathbb{E} = \Psi(u)G(0)\mathbb{F}_1 = r_e(\xi(u))^{-1}\tilde{R}_{\xi(u)}\mathbb{F}_1$ $= \hat{T}(e,\xi(u))\mathbb{F}_1$. Let $\hat{T}^*(\xi(u),\xi(v)) = \Phi_F(\eta(u,v))^{-1}\hat{T}'(\xi(u),\xi(v))\Phi_F(u)$. Then, sliding the index of the fibres from k to $k-r$, we see that $\hat{T}^*$ is equivalent with $\hat{T}'$. Moreover, $\hat{T}^*(\xi(u),\xi(v))\mathbb{F}_1 = \mathbb{F}_1$ by the same reasoning as above. Thus, the image of $\tilde{A}$ is an ILB-subbundle of $B(\mathbb{F}, G, \tilde{T}')$.

Now, to prove the last part of the theorem is to prove that $\hat{T}^*(e,\xi(u))A\hat{T}(e,\xi(u))^{-1}$ satisfies the inequalities of (e). Since

$$\hat{T}^*(e,\xi(u))A\hat{T}(e,\xi(u))^{-1} = \Phi_F(u)^{-1}\hat{T}'(e,\xi(u))A\hat{T}(e,\xi(u))^{-1},$$

we have only to investigate $\Phi_F(u)^{-1}$. Since

$$\Phi_F(u)^{-1}(w_1 + w_2) = w_1 + w_2 - (1 - p)\Psi(u)G(0)I(u)\, w_1 ,$$

and $\|(1 - p)u\|_k \leq C'\|u\|_k + D_k\|u\|_{k-1}$, we have only to show the inequalities for $I(u)$, $(d_1 I)_u$, $(d_1^2 I)_u$, and it is enough to prove the following :

4.3.5 Lemma <u>There is a neighborhood</u> W_3 <u>of</u> 0 <u>in</u> $\mathfrak{g}^{d+r}$ <u>such that the above</u> $I(u)$ <u>satisfies the following inequalities for every</u> $u \in W_3 \cap \mathfrak{g}$:

$$\|I(u)w\|_k \leq C\{\|u\|_k\|w\|_d + \|w\|_k\} + P_k(\|u\|_{k-1})\|w\|_{k-1} ,$$

$$\|(d_1 I)_u(v,w)\|_k \leq C\{\|u\|_k\|v\|_d\|w\|_d + \|v\|_k\|w\|_d + \|v\|_d\|w\|_k\} + P_k(\|u\|_{k-1})\|v\|_{k-1}\|w\|_{k-1} ,$$

$$\begin{aligned}\|(d_1^2 I)_u(v_1,v_2,w)\|_k \leq C\{&\|u\|_k\|v_1\|_d\|v_2\|_d\|w\|_d + \|v_1\|_k\|v_2\|_d\|w\|_d + \|v_1\|_d\|v_2\|_k\|w\|_d \\ &+ \|v_1\|_d\|v_2\|_d\|w\|_k\} \\ &+ P_k(\|u\|_{k-1})\|v_1\|_{k-1}\|v_2\|_{k-1}\|w\|_{k-1} , \quad k \in N(d+r),\ k \geq d+1.\end{aligned}$$

Proof. We put $J(u)w = p\Psi(u)G(0)w$, where w is restricted in $\mathbb{F}_1$. Then, by the inequalities for $\|(d_1\Psi)_u(v,w)\|_{k-r}$ and $\|G(0)v\|_k$, we have

$$\|(dJ)_u(v,w)\|_k \leq C'\{\|u\|_k\|v\|_d\|w\|_d + \|v\|_k\|w\|_d + \|v\|_d\|w\|_k\} + P'_k(\|u\|_{k-1})\|v\|_{k-1}\|w\|_{k-1} .$$

Since $J(0)w = w$ and $J(u)w - w = \int_0^1 (d_1J)_{tu}(u,w)dt$, we have that there exists a neighborhood W' of 0 in $\mathfrak{g}^{d+r}$ such that the following inequality holds for every $u \in W' \cap \mathfrak{g}$:

$$\|J(u)w\|_k \geq \frac{1}{2}\|w\|_k - C''\|u\|_k\|w\|_d - P''_k(\|u\|_{k-1})\|u\|_{k-1}\|w\|_{k-1} .$$

Therefore,

$$\|I(u)w\|_k \leq 2\|w\|_k + C''\|u\|_k\|I(u)w\|_d + P''_k(\|u\|_{k-1})\|u\|_{k-1}\|I(u)w\|_{k-1} .$$

Let W_3 be a neighborhood of 0 in W' such that $\|I(u)w\|_d \leq D\|w\|_d$ for any $u \in W_3$. Then, using the above inequality successively, we obtain

$$\|I(u)w\|_k \leq C\{\|u\|_k\|w\|_d + \|w\|_k\} + P_k(\|u\|_{k-1})\|w\|_{k-1}.$$

Since

$$(d_1I)_u(v,w) = - I(u)(d_1J)_u(v,I(u)w),$$

$$(d_1^2I)_u(v_1,v_2,w) = - (d_1I)_u(v_1,(d_1J)_u(v_2,I(u)w)) - I(u)(d_1^2J)_u(v_1,v_2,I(u)w)$$
$$- I(u)(d_1J)_u(v_2,(d_1I)_u(v_1,w)),$$

we can obtain the desired result by direct computations.

IV.4 Semi-direct product of ILB-Lie groups.

Let $B(\mathbb{F},G,\tilde{T})$ be an ILB-vector bundle defined by the mapping $\tilde{T}$. Assume furthermore that there exists a mapping $\rho : \mathbb{F} \times G \mapsto \mathbb{F}$ such that $\rho(w,gh) = \rho(\rho(w,g),h)$ and that $\tilde{T}(w,g,h)$ is equal to $\rho(w,h)$. Then, ρ has locally the same smoothness property as $\tilde{T}$, but the assumed equality for ρ means that this local property spreads all over the group G. Thus, ρ can be extended to the C^ℓ-mapping of $F^{k+\ell} \times G^k$ into $k \in N(d)$, and hence the local trivialization τ_e is defined for all $g \in G$, that is, we have the mapping $\tau_e : G \times \mathbb{F} \mapsto \mathbb{F} \times G$, $\tau_e(g)w = (\hat{T}(0,g)^{-1}w,g) = (\rho_{g^{-1}}w,g)$, where $\rho_g w$

$= \rho(w,g)$. Therefore, we have that $B(F^k, G^k, \tilde{T})$ is the trivial bundle and the trivialization does not depend on k.

Define the group operation on $\mathbb{F} \times G$ by $(w,g)*(w',g') = (\rho(w,g') + w', gg')$.

4.4.1 Lemma (Cf. Theorem G in § 0.) $\mathbb{F} \times G$ with the above group operation is a strong ILB-Lie group. (We call this a semi-direct product of $\mathbb{F}$ and G, and denote by $\mathbb{F}*G$.)

Proof. We have only to check the properties $(G,1) \sim (G,8)$. $(G,2)$, $(G,3)$ and $(G,8)$ are trivial. $(G,1)$ and $(G,4)$ are easy because the group operation can be extended to the C^ℓ-mapping of $(F^{k+\ell}*G^{k+\ell}) \times (F^k*G^k)$ into F^k*G^k. The inversion is given by $(w,g)^{-1} = (-\rho(w,g^{-1}), g^{-1})$, hence $(G,5)$ is true. $(G,6)$ is trivial. $(G,7)$ is shown by the equality $dR_{(w,g)}(w',u) = (\rho(w',g) + w, dR_g u)$.

Remark The right translation $\tilde{R}(w,g,h)$ in 4.1.2 Lemma is equal to $(w,g)*(0,h)$.

IV.5 Examples of semi-direct products.

Here all arguments are given by using H-norms. However, the results here are also true for B-norms.

Let $\Gamma(1_M)$ be the space of all smooth functions on M (i.e. 1_M means the trivial bundle $R \times M$.) Let $\Gamma^k(1_M)$ be the Hilbert space defined as in § II. By virtue of 2.1.2 Lemma, we have that the mapping $\rho : \Gamma(1_M) \times \mathcal{D} \longmapsto \Gamma(1_M)$ defined by $\rho(f,\varphi)(x) = f(\varphi(x))$ can be extended to the C^ℓ-mapping of $\Gamma^{k+\ell}(1_M) \times \mathcal{D}^k$ into $\Gamma^k(1_M)$ for every $k \geq \dim M + 5$. Thus, we get a semi-direct product $\Gamma(1_M)*\mathcal{D}$.

However, the purpose of this section is to discuss about a slight modification of the above example and to get a strong ILH-Lie group $\Gamma_*(1_M)*\mathcal{D}$ which will be an ambient space of the group of all contact transformations.

Let $\Gamma^k_*(1_M)$ be a subset of all elements f of $\Gamma^k(1_M)$ with $f(x) > 0$ for all $x \in M$ or $f(x) < 0$ for all $x \in M$. We put $\Gamma_*(1_M) = \cap \Gamma^k_*(1_M)$ with the inverse limit topology. $\Gamma^k_*(1_M)$ is defined for $k \geq [\frac{1}{2}\dim M] + 1$ by virtue of Sobolev embedding theorem and an open subset of $\Gamma^k(1_M)$, hence a Hilbert manifold. Moreover,

$\Gamma^k_*(1_M)$ is a Hilbert Lie group for $k \geqslant \dim M + 1$, that is, if $f, f' \in \Gamma^k_*(1_M)$, then $ff' \in \Gamma^k_*(1_M)$. This operation is obviously smooth because this is bi-linear.

Define the group operation $*$ on $\Gamma_*(1_M) \times \mathcal{D}$ by $(f,\varphi)*(f',\varphi') = (\rho(f,\varphi')f',\varphi\varphi')$. Then it is easy to see the following :

4.5.1 Theorem $\Gamma_*(1_M) \times \mathcal{D}$ <u>with the group operation</u> $*$ <u>is a strong ILH-Lie group.</u>

We denote this by $\Gamma_*(1_M)*\mathcal{D}$.

The tangent space of $\Gamma_*(1_M)*\mathcal{D}$ at the identity $(1,e)$ is obviously $\Gamma(1_M) \oplus \Gamma(T_M)$.

4.5.2 Lemma <u>The Lie algebra structure of</u> $\Gamma(1_M) \oplus \Gamma(T_M)$ <u>is given by</u>

$$[(f,u),(g,v)] = (vf - ug,\ [u,v]).$$

Proof. This is obtained by a direct computation of

$$[(f,u),(g,v)] = \frac{\partial^2}{\partial t\partial s}\Big|_{\substack{t=0\\ s=0}}(e^{tf},\exp tu)*(e^{sg},\exp sv)*(e^{tf},\exp tu)^{-1}$$

$$= \left(\frac{\partial^2}{\partial s\partial t}\Big|_{\substack{t=0\\ s=0}} tf(\exp sv \exp -tu) + \frac{\partial^2}{\partial t\partial s}\Big|_{\substack{s=0\\ t=0}} sg(\exp -tu),\ [u,v] \right).$$

Let $\xi : U \cap \Gamma(T_M) \mapsto \tilde{U}$ be the same ILH-coordinate mapping as in II,1. The following is easy to prove by direct computations :

4.5.3 Lemma <u>Define a coordinate mapping</u> $\xi' : \Gamma(1_M) \oplus U \cap \Gamma(T_M) \mapsto \Gamma_*(1_M)*\mathcal{D}$ <u>by</u> $\xi'(f,u) = (e^f, \xi(u))$. <u>Then, we have the following formulae</u> :

$$\eta'((f,u),(g,v)) = \xi'^{-1}(\xi'(f,u)*\xi'(g,v)) = (R'(f,u) + g,\ \eta(u,v))$$

$$\theta'((h,w),(f,u),(g,v)) = (d\eta'_{(g,v)})_{(f,u)}(h,w) = (R'(h,v),\ \theta(w,u,v)),$$

<u>where</u> $R'(f,v)(x) = f(\xi(v)(x))$.

Let K be a compact subgroup of $\mathcal{D}$. As in II.2, we take a smooth K-invariant riemannian metric on M. Then, we see that $A_k(u) = \xi^{-1}(k\xi(u)k^{-1})$ is equal with $Ad(k)u$. (Cf. the proof of 2.2.1.)

Let $A_{(1,k)}(f,u) = \xi'^{-1}((1,k)*\xi'(f,u)*(1,k)^{-1})$,

$$Ad((1,k))(f,u) = \frac{d}{dt}\Big|_{t=o}(1,k)*\xi'(tf,tu)*(1,k)^{-1}.$$

Then, we have $A_{(1,k)}(f,u) = Ad((1,k))(f,u)$. Therefore, by the same manner as in 2.3.2 - 3 Theorems, we have the following :

4.5.4 Theorem <u>Let</u> K <u>be a compact subgroup of</u> $\mathcal{D}$. <u>Then,</u> $G_K = \{(f,\varphi) \in \Gamma_*(1_M)*\mathcal{D} : (1,k)*(f,\varphi) = (f,\varphi)_*(1,k)$ for all $k \in K\}$ <u>is a strong ILH-Lie subgroup of</u> $\Gamma_*(1_M)*\mathcal{D}$.

Let F be a smooth finite dimensional riemannian vector bundle over M, and let $\tilde{T}_F$ be as in IV.2. We put $\tilde{T}'_F(w,(f_1,g_1),(f_2,g_2)) = \tilde{T}_F(w,g_1,g_2)$. $\tilde{T}'_F$ satisfies the conditions (VB, 1 ~ 3) in IV.1 and defines an ILH-vector bundle $B(\Gamma(F),\Gamma_*(1_M)*\mathcal{D},\tilde{T}'_F)$. This is obviously the pull back of $B(\Gamma(F),\mathcal{D},\tilde{T}_F)$ by the natural projection of $\Gamma_*(1_M)*\mathcal{D}$ onto $\mathcal{D}$.

If F is the trivial bundle 1_M of fibre dimension one, then $\tilde{T}_{1_M}(f,g_1,g_2)(x) = R'(f,\xi^{-1}g_2)(x) = \rho(f,g_2)(x) = f(g_2(x))$, hence $\tilde{T}'_{1_M}(f,(f_1,g_1),(f_2,g_2)) = R'(f,\xi^{-1}g_2)$. We denote by $B(\Gamma(1_M), \Gamma_*(1_M)*\mathcal{D}, \tilde{T}'_{1_M})$ the ILH-vector bundle defined by $\tilde{T}'_{1_M}$.

We put

$$\tilde{T}_{1_M \oplus T_M}((f,w),g_1,g_2) = (\tilde{T}_{1_M}(f,g_1,g_2), \tilde{T}_\theta(w,g_1,g_2)),$$

$$\tilde{T}'_\theta((f,w),(f_1,g_1),(f_2,g_2)) = \tilde{T}_{1_M \oplus T_M}((f,w),g_1,g_2).$$ (Cf. 4.1.1, 4.5.3 Lemmas.)

Let $B(\Gamma(1_M) \oplus \Gamma(T_M),\mathcal{D})$, $B(\Gamma(1_M) \oplus \Gamma(T_M), \Gamma_*(1_M)*\mathcal{D})$ be ILH-vector bundles defined by $\tilde{T}_{1_M \oplus T_M}$, $\tilde{T}'_\theta$, respectively. $B(\Gamma(1_M) \oplus \Gamma(T_M),\mathcal{D})$ is the Whitney sum of $B(\Gamma(1_M),\mathcal{D},\tilde{T}_{1_M})$ and $B(\Gamma(T_M),\mathcal{D},\tilde{T}_\theta)$, and $B(\Gamma(1_M) \oplus \Gamma(T_M), \Gamma_*(1_M)*\mathcal{D})$ is the tangent bundle of $\Gamma_*(1_M)*\mathcal{D}$. This is also the Whitney sum of $B(\Gamma(1_M),\Gamma_*(1_M)*\mathcal{D}, \tilde{T}'_{1_M})$ and $B(\Gamma(T_M),\Gamma_*(1_M)*\mathcal{D}, \tilde{T}'_\theta)$ (the pull back of the tangent bundle $B(\Gamma(T_M),\mathcal{D},\tilde{T}_\theta)$).

Now, suppose we have a linear mapping $A : \Gamma(1_M) \oplus \Gamma(T_M) \mapsto \Gamma(F)$ such that A can be extended to a bounded linear mapping of $\Gamma^{k+r}(1_M) \oplus \Gamma^{k+r}(T_M)$ into $\Gamma^k(F)$ for every $k \geq \dim M + 5$. Let A_1, A_2 be the restriction of A onto $\Gamma(1_M)$, $\Gamma(T_M)$ respectively. Then, $A(f,u) = A_1(f) + A_2(u)$. Since A, A_1, A_2 can be regarded as mappings of the fibres at the identity, these can be extended to right invariant bundle morphisms by using the right translations. Thus, we get the following right invariant continuous fibre preserving mappings for every $k \in N(\dim M + 5)$:

$$\tilde{A}' \ : \ B(\Gamma^k(1_M) \oplus \Gamma^k(T_M),\ \Gamma^k_*(1_M)*\mathcal{D}^k) \longrightarrow B(\Gamma^{k-r}(F), \Gamma^k_*(1_M)*\mathcal{D}^k, \tilde{T}'_F)$$

$$\tilde{A} \ : \ B(\Gamma^k(1_M) \oplus \Gamma^k(T_M), \mathcal{D}^k) \longrightarrow B(\Gamma^{k-r}(F),\ \mathcal{D}^k, \tilde{T}_F)$$

$$\tilde{A}'_1 \ : \ B(\Gamma^k(1_M), \Gamma^k_*(1_M)*\mathcal{D}^k,\ \tilde{T}'_{1_M}) \longrightarrow B(\Gamma^{k-r}(F), \Gamma^k_*(1_M)*\mathcal{D}^k,\ \tilde{T}'_F)$$

$$\tilde{A}_1 \ : \ B(\Gamma^k(1_M), \mathcal{D}^k, \tilde{T}_{1_M}) \longrightarrow B(\Gamma^{k-r}(F),\ \mathcal{D}^k,\ \tilde{T}_F)$$

$$\tilde{A}'_2 \ : \ B(\Gamma^k(T_M), \Gamma^k_*(1_M)*\mathcal{D}^k,\ \tilde{T}'_\theta) \longrightarrow B(\Gamma^{k-r}(F), \Gamma^k_*(1_M)*\mathcal{D}^k,\ \tilde{T}'_F)$$

$$\tilde{A}_2 \ : \ B(\Gamma^k(T_M), \mathcal{D}^k,\ \tilde{T}_\theta) \longrightarrow B(\Gamma^{k-r}(F),\ \mathcal{D}^k,\ \tilde{T}_F).$$

Evidently, we have the following relations :

(a) $\tilde{A}'$, $\tilde{A}'_1$, $\tilde{A}'_2$ are pull back of $\tilde{A}$, $\tilde{A}_1$, $\tilde{A}_2$ respectively by the projection $\Gamma_*(1_M)*\mathcal{D} \longrightarrow \mathcal{D}$.

(b) $\tilde{A}(f,u) = \tilde{A}_1(f) + \tilde{A}_2(u)$, $\tilde{A}'(f,u) = \tilde{A}'_1(f) + \tilde{A}'_2(u)$.

Thus, we have the following :

4.5.5 Lemma <u>Notations being as above, if</u> $\tilde{A}_1$ <u>and</u> $\tilde{A}_2$ <u>happen to be smooth bundle morphism, then so is</u> $\tilde{A}'$.

The above lemma will be used later.

§ V Review of the smooth extension theorem and a remark on elliptic operators

V.1 Smooth extension theorems.

In IV.5, we see that the smoothness of $\tilde{A}_i$ implies that of $\tilde{A}'$. However, in general, $\tilde{A}_i$ is not even a continuous bundle morphism. To get the smoothness property, we have to assume (at least at present time) that A_i is a differential operator of order r with smooth coefficients. This was the main theorem of [31]. As we have to use a more detailed fact later as well as the above smooth extension theorem, we will give a short review of [31] and an idea of the proof of it.

In this section, we will give also some inequalities obtained by the smooth extension theorem combined with 2.1.3, 2.5.1 and 2.5.2. These inequalities will be used in the proof of Frobenius theorem. So, one may read the next chapter first to get the idea why we need such inequalities.

Let E and F be smooth riemannian vector bundle over M and let $A : \Gamma(E) \longrightarrow \Gamma(F)$ be a linear differential operator of order r with smooth coefficients.

Recollect the definitions of the ILB- and the ILH- coordinate mapping ξ of $U \cap \Gamma(T_M)$ into $\mathcal{D}$ and of the local trivialization $\tau_e : \tilde{W} \cap \mathcal{D} \times \Gamma(E) \longrightarrow B(\Gamma(E), \mathcal{D}, \tilde{T}_E)$, where τ_e was given by $\tau_e(g)w = (\hat{T}_E(e,g)^{-1}w, g)$. If $E = T_M$, we use the local trivialization $\tau_e(g)w = (\hat{T}_\theta(e,g)^{-1}w, g)$. However, since $B(\Gamma(T_M), \mathcal{D}, \tilde{T}_\theta)$ is the tangent bundle of $\mathcal{D}$, the derivative $d\xi$ of ξ gives also a local trivialization of $B(\Gamma(T_M), \mathcal{D}, \tilde{T}_\theta)$. In general, these two local trivializations coincide. Namely, we have the following :

5.1.1 Lemma Notations being as above, we have $(d\xi)_u w = \tau_e(\xi(u))w$.

Proof. This is only a change of notations. Put $\hat{\theta}(u,v) = \theta(w,u,v)$. Then, we have $\hat{T}_\theta(e,\xi(u))^{-1}w = \hat{\theta}(0,u)^{-1}w$. Recall how the point of $B(\Gamma(T_M), \mathcal{D}, \tilde{T}_\theta)$ is expressed by the pair of elements of $\Gamma(T_M)$ and $\mathcal{D}$. $B(\Gamma(T_M), \mathcal{D}, \tilde{T}_\theta)$ is homeomorphically identified with $\Gamma(T_M) \times \mathcal{D}$ through the mapping $dR : \Gamma(T_M) \times \mathcal{D} \longrightarrow B(\Gamma(T_M), \mathcal{D}, \tilde{T}_\theta)$

defined by $dR(w,g) = dR_g w$. Thus, we have only to show that $\theta(0,u)^{-1}w = dR^{-1}_{\xi(u)}(d\xi)_u w$. Now,

$$\theta(0,u)w = \lim \frac{1}{t} \{\xi^{-1}(\xi(tw)\xi(u) - \xi^{-1}(\xi(u))\} = (dR)_{\xi(u)}w.$$

Since $(d\xi)_u w = w$ by the coordinate expression, we have the desired result.

Remark Instead of the above local trivialization τ_e, we may use the local trivialization τ_e defined by $\widetilde{T}_{T_M}$ as in 4.2.1 in the case $E = T_M$. However, to consider Frobenius theorem, we have to translate distributions on $\mathcal{D}$ into those on $W \cap \Gamma(T_M)$ through the coordinate mapping ξ and at this moment, we have to use $d\xi$ as a local trivialization of the tangent bundle. This is the reason why we use the local trivialization $d\xi$ if $E = T_M$.

Let $\tau'_e : \widetilde{W} \cap \mathcal{D} \longrightarrow B(\Gamma(F), \mathcal{D}, \widetilde{T}_F)$ be the same mapping " τ_e " as above, replacing E by F. By the definition, the local expression of $\widetilde{A}$ is given by $\tau'_e(\xi(u))^{-1}\widetilde{A}\tau_e(\xi(u))w$ and $\tau'_e(\xi(u))^{-1}\widetilde{A}(d\xi)_u w$ if $E = T_M$. These are equal to $(\hat{T}_F(0,\xi(u))A\hat{T}_E(0,\xi(u))^{-1}w,\ u)$ and $(\hat{T}_F(0,\xi(u))A\theta(0,u)^{-1}w,\ u)$ respectively.

Let r be the order of the differential operator A and J^rT_M be the r-th jet bundle of T_M. J^rE means also the r-th jet bundle of E. Let W^r be a relatively compact tubular neighborhood of zero section of J^rT_M. We put

$$\Gamma(W^r) = \{\ u \in \Gamma(T_M) : (j^r u)(x) \in W^r \text{ for every } x \in M\},$$
$$\Gamma^k(W^r) = \{u \in \Gamma^k(T_M) : (j^r u)(x) \in W^r \text{ for any } x \in M\ \},$$
$$\gamma^k(W^r) = \{u \in \gamma^k(T_M) : (j^r u)(x) \in W^r \text{ for any } x \in M\ \},\quad k \geq r,$$

$\Gamma^k(W^r)$ is well-defined for $k \geq [\frac{1}{2}\dim M] + r + 1$ by virtue of Sobolev embedding theorem.

By the arguments in 4°,(b) in [31], we see that for a sufficiently small W^r, there exists a smooth fibre preserving mapping ψ of the closure of W^r into $(J^rE)^* \otimes F$ or $(J^rT_M)^* \otimes F$ such that $\hat{T}_F(0,\xi(u))A\hat{T}_E(0,\xi(u))^{-1}w$ or $\hat{T}_F(0,\xi(u))A(d\xi)_u w$ is defined from ψ, that is, these are equal to $\Psi(j^r u)j^r w$, where $\Psi(j^r u)(j^r w)(x) = \psi(j^r u(x))(j^r w)(x)$, $w \in \Gamma(E)$ or $\Gamma(T_M)$ and $u \in \Gamma(W^r)$.

The explicit expression of ψ is very complicated, using higher order connections defined on jet spaces of mappings. However, this expression has never been used but only the existence of ψ. In fact, even the main theorem of [31] (smooth extension theorem) is an immediate conclusion of this fact combined with 2.1.3 Lemma, because we have the following :

5.1.2 Lemma $j^r : \Gamma(E) \mapsto \Gamma(J^rE)$ can be extended to a bounded linear operator of $\Gamma^{k+r}(E)$ (resp. $\gamma^{k+r}(E)$) into $\Gamma^k(J^rE)$ (resp. $\gamma^k(J^rE)$). Moreover, there is a constant e_r depending only on r such that $\|j^r u\|_k \leq e_r \|u\|_{k+r}$, where $\| \ \|_k$ may be H-norm or B-norm. (Cf. Lemma 16 in [31].)

Apply 2.5.3 Lemma to the mapping $\Psi(j^r u) j^r w$ above, and using 5.1.2, we have the following :

5.1.3 Lemma Put $\Phi(u)w = \Psi(j^r u) j^r w$. Then, the mapping $\Phi : \Gamma(W^r) \times \Gamma(E) \mapsto \Gamma(F)$ can be extended to the smooth mappings $\Phi : \Gamma^k(W^r) \times \Gamma^k(E) \mapsto \Gamma^{k-r}(F)$ for any $k \geq k_1 = \dim M + r + 5$, and $\Phi : \gamma^k(W^r) \times \gamma^k(E) \mapsto \gamma^{k-r}(F)$ for any $k \geq k_1 = r + 1$.

Let $(d_1^s \Phi)_u (u_1, \ldots, u_s, v)$ be the s-times partial derivative of Φ at u with respect to the first variable. Then, for $k \geq k_1 + 1$, and u restricted in a bounded set of $\Gamma^{k_1}(W^r)$ or $\gamma^{k_1}(W^r)$, we have

$$\begin{aligned}
&\|(d_1^s \Phi)_u (u_1, \ldots, u_s, v)\|_{k-r} \\
&\quad \leq C\{ \|u\|_k \|u_1\|_{k_1} \cdots \|u_s\|_{k_1} \|v\|_{k_1} + \sum_{j=1}^{s} \|u_1\|_{k_1} \cdots \|u_j\|_k \cdots \|u_s\|_{k_1} \|v\|_{k_1} \\
&\qquad\qquad + \|u_1\|_{k_1} \cdots \|u_s\|_{k_1} \|v\|_k \} \\
&\qquad + P_k(\|u\|_{k-1}) \|u_1\|_{k-1} \cdots \|u_s\|_{k-1} \|v\|_{k-1} ,
\end{aligned}$$

where C is a positive constant independent from k and P_k is a polinomial with positive coefficients.

Recall the definition of $\tilde{A}$ and its local expression, and we have

5.1.4 Corollary $\tilde{A}$ is a C^∞ILHC^2- and a C^∞ILBC^2- normal bundle morphism.

Now, we back to the setting in IV.5 of the previous chapter and assume we have a linear differential operator $A : \Gamma(1_M) \oplus \Gamma(T_M) \longmapsto \Gamma(F)$ of order r with smooth coefficients. Let A_i, ($i = 1,2$) be as in p.IV.15. By the smooth extension theorem (5.1.3), we see that $\tilde{A}'$ can be extended to a smooth bundle morphism of $B(\Gamma^k(1_M) \oplus \Gamma^k(T_M), \Gamma^k_*(1_M)*\mathcal{D}^k)$ into $B(\Gamma^{k-r}(F), \Gamma^k_*(1_M)*\mathcal{D}^k, \tilde{T}'_F)$ for any $k \geq \dim M + r + 5$. Here, we want to prove that $\tilde{A}'$ is a C^∞ILHC^2-normal bundle morphism.

The local expression of $\tilde{A}'$ is given by $\hat{T}'_F(0,\xi'(f,u))A\hat{\theta}'(0,(f,u))^{-1}(h,w)$. On the other hand,

$$\hat{\theta}'(0,(f,u))^{-1} = \hat{\theta}'((f,u),\iota(f,u)) = \hat{\theta}'((f,u),(-f(\xi(u)),\iota(u)).$$

Thus, by 4.5.3 Lemma, $\hat{\theta}'(0,(f,u))^{-1}(h,w) = (R'(h,\iota(u)), \hat{\theta}(0,u)^{-1}w)$. Therefore,

$$\hat{T}'_F(0,\xi'(f,u))A\hat{\theta}'(0,(f,u))^{-1}(h,w)$$

$$= \hat{T}_F(0,\xi(u))A_1R'(h,\iota(u)) + \hat{T}_F(0,\xi(u))A_2\hat{\theta}(0,u)^{-1}w$$

$$= \hat{T}_F(0,\xi(u))A_1\hat{T}_{1_M}(0,\xi(u))^{-1}h + \hat{T}_F(0,\xi(u))A_2\hat{\theta}(0,u)^{-1}w.$$

Since $\tilde{A}_1$ and $\tilde{A}_2$ are C^∞ILHC^2-normal bundle morphisms, we have the following :

5.1.5 Lemma <u>If</u> A_1, A_2 <u>are differential operators of order</u> r <u>with smooth coefficients, then</u> $\tilde{A}'$ <u>is a</u> C^∞<u>ILHC</u>^{2}<u>-normal bundle morphism.</u>

V.2 Elliptic differential operators.

Recall the conditions (a) ~ (e) in IV,3. The conditions (a) and (e) are satisfied, if A is a linear differential operator with smooth coefficients. So, here we will discuss about the conditions (b) ~ (d).

In this section, we assume always that M is oriented, and we use only H-norms. Theorems in this section can not be true for B-norms.

Let E be a finite dimensional smooth vector bundle over M. Suppose

$$D : \Gamma(E) \longmapsto \Gamma(E)$$

is a linear differential operator of order m with smooth coefficients.

5.2.1 Lemma *If* D *is elliptic, then the following inequality holds for* $s \geq 0$:

$$\|Du\|_s \geq C\|u\|_{s+m} - D_s\|u\|_{s+m-1},$$

where C, D_s *are positive constants such that* C *is independent from* s.

Of course, this is the usual Gårding's inequality. However, the independence of C from s is essential in our case. The precise proof is seen in Lemma 4 in [32].

Let E, F and H be smooth finite dimensional riemannian vector bundles over M. Let $A : \Gamma(E) \mapsto \Gamma(F)$, $B : \Gamma(F) \mapsto \Gamma(H)$ be linear differential operatore with smooth coefficients. Let A^*, B^* be the formal adjoint operators of A, B respectively.

The purpose of this section is to prove the following :

5.2.2 Theorem *Let* A, B *be linear differential operator of order* r *with smooth coefficients. Suppose* $BA = 0$ *and* $AA^* + B^*B$ *is elliptic. Then, the right invariant mapping* $\tilde{A} : B(\Gamma(E), \mathcal{D}, \tilde{T}_E) \longrightarrow B(\Gamma(F), \mathcal{D}, \tilde{T}_F)$ *satisfies the conditions* (a) ~ (e) *in IV.3. Therefore, 4.3.1 Theorem holds for* $\tilde{A}$.

Proof. Recall 5.1.3. We have only to show (b) ~ (d).

Let $\mathfrak{h} = \mathrm{Ker}A^* \cap \mathrm{Ker}B$. Then, $\mathfrak{h}$ is equal to the kernel of $\square = AA^* + B^*B$, hence $\mathfrak{h} \subset \Gamma(F)$ and of finite dimension. Let $\Gamma_\square(F) = \{u \in \Gamma(F) : \int_M < u, e > dV = 0$ for all $e \in \mathfrak{h}\}$, and let $\Gamma^k_\square(F)$ be the closure of $\Gamma_\square(F)$ in $\Gamma^k(F)$. Then,

$$\Gamma(F) = \Gamma_\square(F) \oplus \mathfrak{h}, \qquad \Gamma^k(F) = \Gamma^k_\square(F) \oplus \mathfrak{h},$$

and $\square : \Gamma^{k+2r}_\square(F) \mapsto \Gamma^k_\square(F)$, $k \geq 0$, is an isomorphism. We denote $\square^{-1}$ the inverse of this isomorphism.

Let $\mathbb{F}_1 = \mathrm{Ker}B \cap \Gamma_\square(F)$, $\mathbb{F}_2 = \mathrm{Ker}A^* \cap \Gamma_\square(F)$ and let F^k_i be the closure of $\mathbb{F}_i$ in $\Gamma^k_\square(F)$. Then, Lemma 6 in [31] shows that $\Gamma_\square(F) = \mathbb{F}_1 \oplus \mathbb{F}_2$ and $\Gamma^k_\square(F) = F^k_1 \oplus F^k_2$ (ILH-splitting). Moreover, $F^k_1 = AA^*F^{k+2r}_1$, $F^k_2 = B^*BF^{k+2r}_2$, and the mapping $\square : F^{k+2r}_i \longrightarrow F^k_i$ is an isomorphism.

Let $\mathbb{E}_1 = \mathrm{Ker}\{A : \Gamma(E) \mapsto \Gamma(F)\}$ and $\mathbb{E}_2 = A^*\Gamma(F)$. Let E^k_i, $i = 1,2$, be the closure of $\mathbb{E}_i$ in $\Gamma^k(E)$. Since $AA^* : F^{k+2r}_1 \longrightarrow F^k_1$ is an isomorphism and

$A^*\Gamma^{k+2r}(F) = A^*F_1^{k+2r}$, we see that $A^*\Gamma^{k+2r}(F)$ is closed in $\Gamma^k(E)$ and hence equal to E_2^k.

For any $u \in \mathbb{\Gamma}(E)$, the element $u - A^*\Box^{-1}Au$ is contained in $\mathbb{E}_1$. Thus, the identity $u = (u - A^*\Box^{-1}Au) + A^*\Box^{-1}Au$ gives the ILH-splitting $\mathbb{\Gamma}(E) = \mathbb{E}_1 \oplus \mathbb{E}_2$, $\Gamma^k(E) = E_1^k \oplus E_2^k$. Thus, the condition (b) is satisfied.

Let $p : \mathbb{\Gamma}(F) \mapsto \mathbb{F}_1$ be the projection. Then, p is given by $\Box^{-1}AA^*$. (Cf. Corollary 2 [31].) Since AA^* is a differential operator of order $2r$ with smooth coefficients, we see that there is a smooth section α of $(J^{2r}F)^* \otimes F$ such that $AA^*u = \alpha j^{2r}u$. Apply 2.5.1 Lemma and 5.1.2 Lemma to the operator αj^{2r}, and we get $\|AA^*u\|_{k-2r} \leq C'\|u\|_k + D_k'\|u\|_{k-1}$.

On the other hand, 5.2.1 Lemma shows that

$$\|\Box^{-1}u\|_k \leq C''\|u\|_{k-2r} + D_k''\|u\|_{k-2r-1},$$

hence $\|pu\|_k \leq C\|u\|_k + D_k\|u\|_{k-1}$. The condition (c) is satisfied.

Consider the mapping $A : \mathbb{E}_2 \mapsto \mathbb{F}_1$. Since $\mathbb{E}_2 = A^*\mathbb{F}_1$, we see that $Au = AA^*v = \Box v$, $v \in \mathbb{F}_1$. Thus,

$$\|Au\|_{k-r} = \|\Box v\|_{k-r} \geq C'\|v\|_{k+r} - D_k'\|v\|_{k+r-1}.$$

On the other hand, $\|A^*v\|_k \leq C''\|v\|_{k+r} + D_k''\|v\|_{k+r-1}$. Therefore, we have

$$\|Au\|_{k-r} \geq C\|u\|_k - D_k'\|v\|_{k+r-1}.$$

Since $A^* : F_1^{k+r-1} \mapsto E_2^{k-1}$ is an isomorphism, we see $\|u\|_{k-1} = \|A^*v\|_{k-1} \geq C_{k-1}\|v\|_{k+r-1}$. Therefore,

$$\|Au\|_{k-r} \geq C\|u\|_k - D_k\|u\|_{k-1}.$$

The condition (d) is satisfied.

Now, recall the formula of the local expression of the right invariant C^∞ILHC^2-normal bundle morphism $\tilde{A}' : B(\mathbb{\Gamma}(1_M) \oplus \mathbb{\Gamma}(T_M), \mathbb{\Gamma}_*(1_M)*\mathfrak{D}) \mapsto B(\mathbb{\Gamma}(F), \mathbb{\Gamma}_*(1_M)*\mathfrak{D}, \tilde{T}_F')$. (See just above the 4.5.5 Lemma.) Then, we have the following :

5.2.3 Theorem <u>Let</u> $A : \mathbb{\Gamma}(1_M) \oplus \mathbb{\Gamma}(T_M) \mapsto \mathbb{\Gamma}(F)$, $B : \mathbb{\Gamma}(F) \mapsto \mathbb{\Gamma}(H)$ <u>be linear differential operators with smooth coefficients such that</u> $BA = 0$ <u>and</u> $\Box = AA^* + B^*B$ <u>is elliptic</u>.

<u>Then,</u> $\widetilde{A}'$ <u>satisfies the conditions</u> (a) ~ (e). <u>Therefore, 4.3.1 Theorem holds for</u> $\widetilde{A}'$.

The above theorem will be used in the proof that the group of all contact transformations is a strong ILH-Lie subgroup of $\Gamma_*(1_M)*\mathcal{D}$.

§ VI Basic theorems II (Frobenius theorem)

VI.1 Basic idea for Frobenius theorem.

The main part of this section has been already discussed in 1°, (b) in [32]. However, the results here are slightly stronger and these are key point of this article. So, these will be repeated in this section.

Let G be a strong ILB-Lie group with the Lie algebra $\mathfrak{g}$. Let $\mathfrak{g}^k$ be the tangent space of G^k at the identity. Roughly speaking, many local problems on G can be translated into those on an open subset $U \cap \mathfrak{g}$ through an ILB-coordinate (U,ξ), where U is an open subset of 0 of $\mathfrak{g}^d$.

Here, we start with a Sobolev chain $\{\mathfrak{g}, \mathfrak{g}^k, k \in N(d)\}$ and assume the following :

(A) There exist closed subspaces S^k and T^k of $\mathfrak{g}^k$ for every $k \geq d$, and $\mathbf{S}$, $\mathbf{T}$ of $\mathfrak{g}$ such that $\mathfrak{g} = \mathbf{S} \oplus \mathbf{T}$, $\mathfrak{g}^k = S^k \oplus T^k$ and $\{\mathbf{S}, S^k, k \in N(d)\}$, $\{\mathbf{T}, T^k, k \in N(d)\}$ are Sobolev chains. Moreover, every T^k is a Hilbert space and $\mathfrak{g}^k$ has the norm $\| \ \|_k$ defined by $\|u+v\|_k^2 = \|u\|_k^2 + \|v\|_k^2$, $u \in \mathbf{S}$, $v \in \mathbf{T}$.

Remark If $\mathfrak{g} = \mathbf{S} \oplus \mathbf{T}$ is an ILB-normal splitting, then we can always **replace** the norm $\| \ \|_k$ on $\mathfrak{g}^k$ by $(\|u\|_k^2 + \|v\|_k^2)^{\frac{1}{2}}$. Nothing will be chaged in the argument of this article.

Let U be an open neighborhood of 0 of $\mathfrak{g}^d$. We may assume without loss of generality that there exist open neighborhoods V, W of zeros in S^d, T^d respectively such that $U = V \oplus W$ (direct product). Assume furthermore that there exists a mapping Φ of $U \cap \mathfrak{g} \times \mathbf{S}$ into $\mathbf{T}$ satisfying the following :

(i) $\Phi(0)v = 0$ and $\Phi(u)v$ is linear with respect to the second variable v.
(ii) Φ can be extended to the smooth mapping of $(U \cap \mathfrak{g}^k) \times S^k$ into T^k for every $k \in N(d)$.
(iii) Let $D_u^k = \{(v,\Phi(u)v) : v \in S^k\}$. Then, $D^k = \{ D_u^k : u \in U \cap \mathfrak{g}^k\}$ gives a smooth

involutive distribution on $U \cap \mathfrak{J}^k$.

Roughly speakeing, the local problem about Frobenius theorem begins with setting this situation mentioned above. Of course, there are several problems in changing the given distribution to the above local situation. However, different from the case of Banach manifolds, such setting of problems does not necessarily imply the existence of integral submanifolds.

Anyhow, we assume the following inequality (✪) :

$$(\text{✪}) \quad \|\Phi(u)v\|_k \leq C_k\{\|u\|_k\|v\|_d + \|v\|_k\} + \gamma_k(\|u\|_{k-1})\|v\|_{k-1}, \quad k \geq d+1,$$

where C_k is a positive constant which may or may not depend on k and γ_k is a positive continuous function.

Now, as in [8] p305, consider the following equation :

$$(\mathrm{E}) \qquad \frac{d}{dt}y(t) = \Phi(tx + y(t))x, \quad x \in V \cap \mathbf{S}, \quad y(t) \in W \cap \mathbf{T}.$$

By the above condition (ii), this equation can be regarded as that on $U \cap \mathfrak{J}^k$ for every k. Assume for a moment that $x \in V \cap S^k$ and $y(0) \in W \cap T^k$ (resp. $x \in V \cap \mathbf{S}$, $y(0) \in W \cap \mathbf{T}$). Then, for any s such that $s \leq k$ (resp. $s < \infty$), there exists $t_s > 0$ such that $y(t)$, $0 \leq t < t_s$, is the solution of (E) in $W \cap T^s$ with the initial condition $y(0)$. Assume t_s is the maximal number in such t_s. By the condition (ii) again, we see easily that $t_s \leq t_{s-1}$. However, we can not conclude $t_s = t_{s-1}$ without the inequality (✪). (If $\lim_{s\to\infty} t_s = 0$, then there is no solution in $W \cap \mathfrak{J}$.)

6.1.1 Lemma (regularity of solutions) Let $x \in V \cap S^k$, $y(0) \in W \cap T^k$ (resp. $x \in V \cap \mathbf{S}$, $y(0) \in W \cap \mathbf{T}$). Assume $y(t)$ is the solution of (E) in W with the initial condition $y(0)$. If Φ satisfies (✪), then $y(t)$ is contained in $W \cap T^k$ (resp. in $W \cap \mathbf{T}$).

Proof. Assume $t_d = \cdots = t_{s-1} > t_s$. Then, obviously $y(t_s) \notin W \cap T^s$, while $y(t_s) \in W \cap T^{s-1}$. If $\|y(t)\|_s$ is bounded in $t \in [0, t_s)$, then so does $\|\frac{d}{dt}y(t)\|_s$ by virtue of the equality (E) and the inequality (✪). Hence $\lim_{t\to t_s} y(t)$ exists in $W \cap T^s$.

Therefore, we can extend the solution beyond t_s. This is contradiction. Thus, $\|y(t)\|_s$ is unbounded.

On the other hand, since the norm on T^k is differentiable, we see

$$\frac{d}{dt}\|y(t)\|_s^2 \leq 2\|\Phi(tx + y(t))x\|_s\|y(t)\|_s \leq \|\Phi(tx + y(t)\|_s^2 + \|y(t)\|_s^2 .$$

Since $\|y(t)\|_{s-1}$ is bounded, using the inequality (✪), we see

$$\frac{d}{dt}\|y(t)\|_s^2 \leq C''\|y(t)\|_s^2 + K'_s .$$

Hence, $\|y(t)\|_s^2$ is not larger than the solution of $\frac{d}{dt}f(t) = C''_k f(t) + K'_s$. Thus, $\|y(t)\|_s^2$ is bounded in $[0,t_s)$. Therefore, we have $t_s = t_{s-1}$. This implies that if $y(t)$ is the solution in W, then so is $y(t)$ in $W \cap T^s$ (resp. $W \cap \mathbf{T}$).

Consider the equation

$$(E^{-1}) \quad \frac{d}{dt}y(1-t) = -\Phi((1-t)x + y(1-t))x,$$

and we have the following :

6.1.2 Corollary Let $y(t)$ be a solution of (E) in W. Assume $x \in V \cap S^k$ and $y(1) \in W \cap T^k$ (resp. $x \in V \cap \mathbf{S}$, $y(1) \in W \cap \mathbf{T}$). Then, $y(t)$ is contained in $W \cap T^k$ (resp. $W \cap \mathbf{T}$).

We keep the notations as above. Since D^d is a smooth involutive distribution on U, Frobenius theorem on Banach manifolds [8] shows that there exist open star shaped neighborhoods V_1, W_1 of 0 in V, W respectively and a smooth diffeomorphism Ψ of $V_1 \oplus W_1$ onto an open neighborhood of 0 in $V \oplus W$ such that $\Psi(V_1, w)$ is an integral submanifold of the involutive distribution D^d through w. This mapping Ψ is given by the following manner : Let $\bar{y}(x,y,t)$ be the solution of (E) with the initial condition y. Then, Ψ is given by $\Psi(x,y) = (x,\bar{y}(x,y,1))$.

Thus, by 6.1.1 Lemma, Ψ is defined as a mapping of $V_1 \cap S^k \oplus W_1 \cap T^k$ into $U \cap \mathfrak{g}^k$ and by the differential dependency on the initial conditions, we see that Ψ is a smooth mapping of $V_1 \cap S^k \oplus W_1 \cap T^k$ into $U \cap \mathfrak{g}^k$ for every $k \in N(d)$. Obviously, $\Psi : V_1 \cap S^k \oplus W_1 \cap T^k \longrightarrow U \cap \mathfrak{g}^k$ is injective, and 6.1.2 Corollary shows

$\Psi(V_1 \cap S^k, W_1 \cap T^k) = \Psi(V_1, W_1) \cap \mathfrak{g}^k$.

6.1.3 Proposition Ψ <u>is a smooth diffeomorphism of</u> $V_1 \cap S^k \oplus W_1 \cap T^k$ <u>onto</u> $\Psi(V_1, W_1) \cap \mathfrak{g}^k$ <u>for every</u> $k \in N(d)$. <u>Moreover,</u> $\Psi(V_1 \cap S^k, w)$ <u>is an integral submanifold of</u> D^k <u>through</u> w <u>for every</u> $w \in W_1 \cap T^k$.

Proof. It is enough to prove that the derivative $(d\Psi)_{(x,y)}$ at (x,y) is an isomorphism of $\mathfrak{g}^k$ onto itself. Since $(d\Psi)_{(x,y)}$ is injective, we have only to prove the surjectivity. The derivative is given by

$$(d\Psi)_{(x,y)} = \begin{pmatrix} \text{id.}, & 0, \\ (d_1\bar{y})_{(x,y,1)}, & (d_2\bar{y})_{(x,y,1)} \end{pmatrix}$$

where $(d_i\bar{y})_{(x,y,1)}$ is the partial derivative of $\bar{y}$ at $(x,y,1)$ with respect to the i-th variable. Therefore, we have only to show that $(d_2\bar{y})_{(x,y,1)}$ is surjective.

Put $z(t) = (d_2\bar{y})_{(x,y,t)}z$. Then, $z(t)$ satisfies the equation

$$\frac{d}{dt} z(t) = (d_1\Phi)_{\Psi(tx,y)}(z(t),x).$$

Since the above equation is linear and x, y are fixed, we see that for any given $z(1)$, we can find the solution $z(t)$. Put $z(0) = z$. Then, $z(1) = (d_2\bar{y})_{(x,y,1)}z$. Thus, $d\Psi$ is an isomorphism.

VI.2 A sufficient condition which ensures that Ψ is a $C^\infty ILBC^2$-normal mapping.

In the above proposition, we gave a basic idea for Frobenius theorem. However, it would be convenient, if we could obtain stronger properties for the resulting ILB-coordinate mapping Ψ. Namely, sometimes, we need that Ψ is a $C^\infty ILBC^2$-normal mapping. So the purpose of this cection is to give a sufficient condition for that.

6.2.1 Theorem <u>Keep the notations and assumptions as in VI.1. Assume furthermore that</u> Φ <u>satisfies the same inequalities as in</u> p.IV.6, (e) <u>replacing</u> Ψ <u>by</u> Φ <u>and puttin</u> $r = 0$, (<u>we may call</u> Φ <u>a</u> $C^\infty ILBC^2$<u>-normal bundle morphism of order</u> 0,) <u>then the resulting ILB-coordinate mapping</u> Ψ <u>in 6.1.3 Proposition is a</u> $C^\infty ILBC^2$<u>-normal mapping</u>

<u>such that</u> $(d\Psi)_{(0,0)} = \mathrm{id.}$. (<u>Therefore, we can use the implicit function theorem.</u>)

Proof. Recall the conditions of $C^\infty ILBC^2$-normal mappings. (Cf. 3.1.5.) Ψ is defined by $\Psi(x,y) = (x,\varphi(x,y))$, $\varphi(x,y) = \bar{y}(x,y,1)$, and $\mathfrak{g} = \mathbb{S} \oplus \mathbb{T}$ is an ILB-normal splitting, we have only to show that φ is a $C^\infty ILBC^2$-normal mapping.

Let $\bar{y}(x,y,t)$ be the solution of (E) with the initial condition $\bar{y}(x,y,0) = y$. We may assume for sufficiently small bounded neighborhoods V_1, W_1 of zeros of S^d, T^d that the following inequalities hold for every $x \in V_1$, $y \in W_1$:

$$\|(d\bar{y})_{(x,y,t)}(x_1,y_1)\|_d \leq K\|(x_1,y_1)\|_d$$

$$\|(d^2\bar{y})_{(x,y,t)}((x_1,y_1),(x_2,y_2))\|_d \leq K\|(x_1,y_1)\|_d\|(x_2,y_2)\|_d , \quad t \in [0,1].$$

First of all, we want to prove the inequality

$$\|\bar{y}(x,y,t)\|_k \leq C(\|x\|_k + \|y\|_k) + P_k(\|x\|_{k-1} + \|y\|_{k-1})$$

by the induction with respect to the index k. Since $\bar{y}(o,o,t) = 0$, and $\bar{y}(x,y,t) = \int_0^1 (d\bar{y})_{(sx,sy,t)}(x,y)ds$, we see that $\|\bar{y}(x,y,t)\|_d \leq K\|(x,y)\|_d$. So, the first step of the induction is established. Assume the above inequality hold for $\leq k-1$. $\bar{y}(x,y,t)$ satisfies the equation $\frac{d}{dt}\bar{y}(x,y,t) = \Phi(tx + \bar{y}(x,y,t))x$, $t \in [0,1]$, so that

$$\frac{d}{dt}\|\bar{y}(x,y,t)\|_k^2 \leq \|\Phi(tx + \bar{y}(x,y,t))x\|_k^2 + \|\bar{y}(x,y,t)\|_k^2$$

$$\leq C'\{\|tx + \bar{y}(x,y,t)\|_k^2\|x\|_d^2 + \|x\|_k^2\} + \|\bar{y}(x,y,t)\|_k^2 + P_k'(\|tx\|_{k-1}^2 + \|\bar{y}(x,y,t)\|_{k-1}^2)\|x\|_{k-1}^2.$$

Since $\|x\|_d$ is bounded, we have

$$\frac{d}{dt}\|\bar{y}(x,y,t)\|_k^2 \leq C''\{\|x\|_k^2 + \|\bar{y}(x,y,t)\|_k^2\} + P_k''(\|x\|_{k-1}^2 + \|y\|_{k-1}^2)$$

by using the assumption of the induction. Hence,

$$\|\bar{y}(x,y,t)\|_k^2 \leq \|y\|_k^2 e^{C''t} + D\{\|x\|_k^2 + P_k''(\|x\|_{k-1}^2 + \|y\|_{k-1}^2)\} e^{C''t}.$$

Therefore, $\|\bar{y}(x,y,t)\|_k \leq C\{\|x\|_k + \|y\|_k\} + P_k(\|x\|_{k-1}+\|y\|_{k-1})$. This is the desired inequality.

Let $(d\bar{y})_{(x,y,t)}(x_1,y_1)$ be the derivative of $\bar{y}$ at (x,y,t) with respect to the variables (x,y). We want to prove next that $(d\bar{y})$ satisfies the inequality

$$\|(d\bar{y})_{(x,y,t)}(x_1,y_1)\|_k \leq C\{(\|x\|_k + \|y\|_k)(\|x_1\|_d + \|y_1\|_d) + \|x_1\|_k + \|y_1\|_k\}$$
$$+ P_k(\|x\|_{k-1} + \|y\|_{k-1})(\|x_1\|_{k-1} + \|y_1\|_{k-1}).$$

We use induction again. Since $\|(d\bar{y})_{(x,y,t)}(x_1,y_1)\|_d \leq K\|(x_1,y_1)\|_d$, the first step of the induction is given. So, we assume the above inequality holds for $\leq k-1$. Now, $(d\bar{y})_{(x,y,t)}(x_1,y_1)$ satisfies the equation

$$\frac{d}{dt}(d\bar{y})_{(x,y,t)}(x_1,y_1) = (d_1\Phi)_{tx+\bar{y}(x,y,t)}(tx_1 + (d\bar{y})_{(x,y,t)}(x_1,y_1),\ x)$$
$$+ \Phi(tx + \bar{y}(x,y,t))x_1, \qquad t \in [0,1],$$

so that using the above result and the assumption of the induction,

$$\frac{d}{dt}\|(d\bar{y})_{(x,y,t)}(x_1,y_1)\|_k^2 \leq C'\{\|(d\bar{y})_{(x,y,t)}(x_1,y_1)\|_k^2 + \|x_1\|_k^2\}$$
$$+ D'(\|x\|_k^2 + \|y\|_k^2)(\|x_1\|_d^2 + \|y_1\|_d^2)$$
$$+ P_k'(\|x\|_{k-1}^2 + \|y\|_{k-1}^2)(\|x_1\|_{k-1}^2 + \|y_1\|_{k-1}^2).$$

Remark that $\bar{y}(x,y,0) = y$, hence $(d\bar{y})_{(x,y,0)}(x_1,y_1) = y_1$. So the above inequality shows that

$$\|(d\bar{y})_{(x,y,t)}(x_1,y_1)\|_k^2 \leq C''\{(\|x\|_k^2 + \|y\|_k^2)(\|x_1\|_d^2 + \|y_1\|_d^2) + \|x_1\|_k^2 + \|y_1\|_k^2\}$$
$$+ P_k''(\|x\|_{k-1}^2 + \|y\|_{k-1}^2)(\|x_1\|_{k-1}^2 + \|y_1\|_{k-1}^2).$$

It follows the desired inequality.

At the end of the proof, we have to show the following inequality :

$$\|(d^2\bar{y})_{(x,y,t)}((x_1,y_1),(x_2,y_2))\|_k \leq C\{(\|x\|_k + \|y\|_k)(\|x_1\|_d + \|y_1\|_d)(\|x_2\|_d + \|y_2\|_d)$$
$$+ (\|x_1\|_d + \|y_1\|_d)(\|x_2\|_k + \|y_2\|_k) + (\|x_1\|_k + \|y_1\|_k)(\|x_2\|_d + \|y_2\|_d)\}$$
$$+ P_k(\|x\|_{k-1} + \|y\|_{k-1})(\|x_1\|_{k-1} + \|y_1\|_{k-1})(\|x_2\|_{k-1} + \|y_2\|_{k-1}).$$

We use induction again. Remark that $d^2\bar{y}$ satisfies the equation

$$\frac{d}{dt}(d^2\bar{y})_{(x,y,t)}((x_1,y_1),(x_2,y_2))$$
$$= (d_1\Phi)_{tx+\bar{y}(x,y,t)}((d^2\bar{y})_{(x,y,t)}((x_1,y_1),(x_2,y_2)),\ x)$$
$$+ (d_1\Phi)_{tx+\bar{y}(x,y,t)}(tx_1 + (d\bar{y})_{(x,y,t)}(x_1,y_1),\ x_2)$$
$$+ (d_1^2\Phi)_{tx+\bar{y}(x,y,t)}(tx_1 + (d\bar{y})_{(x,y,t)}(x_1,y_1),\ tx_2 + (d\bar{y})_{(x,y,t)}(x_2,y_2),\ x)$$
$$+ (d_1\Phi)_{tx+\bar{y}(x,y,t)}(tx_2 + (d\bar{y})_{(x,y,t)}(x_2,y_2),\ x_1).$$

Thus, using the assumption for Φ and the assumption of the induction, we get the desired result after a little bit complicated computation. There is no technical difficulty. So, we would like to omit the precise proof.

Now, the theorem is proved , because $\varphi(x,y) = \bar{y}(x,y,1)$.

VI.3 Distributions defined by the kernel of C^∞ILBC^2-normal bundle morphisms

The assumptions imposed on Φ in the previous sections are fairly strong. So, it is hard to apply the above result directly. Here, we will give a sufficient condition under which the conditions (i ~ iii) and the conditions of 6.2.2 Theorem are satisfied.

Now, we back to the situation of 4.3.1 Theorem and consider a right invariant bundle morphism $\tilde{A} : B(\mathfrak{g}, G, \tilde{T}_\theta) \rightarrow B(\mathbb{F}, G, \tilde{T}')$ satisfying the conditions (a ~e) there, where $B(\mathfrak{g}, G, \tilde{T}_\theta)$ is the tangent bundle of a strong ILB-Lie group G and $\tilde{T}_\theta(w,g,h) = \theta(w,\xi^{-1}(g),\xi^{-1}(h))$. The ILB-splitting in the condition (b) will be denoted by $\mathfrak{g} = \mathbb{E}_1 \oplus \mathbb{E}_2$, $\mathbb{F} = \mathbb{F}_1 \oplus \mathbb{F}_2$. First of all we should remark the following :

6.3.1 Lemma <u>Notations and assumptions being as above, the splittings</u> $\mathfrak{g} = \mathbb{E}_1 \oplus \mathbb{E}_2$, $\mathbb{F} = \mathbb{F}_1 \oplus \mathbb{F}_2$ <u>are in fact ILB-normal splittings</u>.

Proof. Since the projection $p : \mathbb{F} \mapsto \mathbb{F}_1$ satisfies the inequality $\|pu\|_k \leq C\|u\|_k + D_k\|u\|_{k-1}$, it is easy to see that $\mathbb{F} = \mathbb{F}_1 \oplus \mathbb{F}_2$ is an ILB-normal splitting.

Now, consider the mapping $A : \mathfrak{g} = \mathbb{E}_1 \oplus \mathbb{E}_2 \mapsto \mathbb{F} = \mathbb{F}_1 \oplus \mathbb{F}_2$. $\mathbb{E}_1$ is the kernel of A and $A : E_2^{k+r} \mapsto F_1^k$ is an isomorphism. Denote by G its inverse. The splitting

$\mathfrak{g} = \mathbb{E}_1 \oplus \mathbb{E}_2$ is given by $u = (u - GAu) + GAu$. Thus, we have only to show that

$$\|GAu\|_k \leqslant C\|u\|_k + D_k\|u\|_{k-1}, \quad k \geqslant d+r+1.$$

This inequality follows immediately from the inequalities

$\|Gu\|_k \leqslant C\|u\|_{k-r} + D_k\|u\|_{k-r-1}$ (cf. the condition (d)) and

$\|Au\|_{k-r} \leqslant C\|u\|_k + D_k\|u\|_{k-1}$. (This is obtained by putting $u = 0$ in the condition (e).)

The goal of this section is to prove the following :

6.3.2 Theorem Let $\widetilde{A} : B(\mathfrak{g}, G, \widetilde{T}_\theta) \longmapsto B(\mathbb{F}, G, \widetilde{T}')$ be a right invariant C^∞ILBC^2-normal bundle morphism of order r satisfying the conditions (b) ~ (d) in IV.3. Then, there are an open neighborhood W of 0 in $\mathfrak{g}^{d+r}$ and a mapping $\Phi : W \cap \mathfrak{g} \times \mathbb{E}_1 \mapsto \mathbb{E}_2$ satisfying the following :

(i) $\Phi(0)v = 0$, and $\Phi(u)v$ is linear with respect to the second variable v.

(ii) Φ can be extended to a smooth mapping of $W \cap \mathfrak{g}^k \times \mathbb{E}_1^k$ into $\mathbb{E}_2^k$ for every $k \in N(d+r)$.

(iii) $(d\xi)_u(\{(v, \Phi(u)v) : v \in \mathbb{E}_1^k\})$ is the kernel of $\widetilde{A} : B(\mathfrak{g}^k, G^k, \widetilde{T}_\theta) \longrightarrow B(\mathbb{F}^{k-r}, G^k, \widetilde{T}')$ at the fibre on $\xi(u)$ for every $k \in N(d+r)$.

(iv) Φ satisfies the following inequalities for $k \in N(d+r)$:

$$\|\Phi(u)w\|_k \leqslant C\{\|u\|_k\|w\|_d + \|v\|_k\} + P_k(\|u\|_{k-1})\|w\|_{k-1} ,$$

$$\|(d_1\Phi)_u(v,w)\|_k \leqslant C\{\|u\|_k\|v\|_d\|w\|_d + \|v\|_k\|w\|_d + \|v\|_d\|w\|_k\} + P_k(\|u\|_{k-1})\|v\|_{k-1}\|w\|_{k-1} ,$$

$$\|(d_1^2\Phi)_u(v_1,v_2,w)\|_k \leqslant C\{\|u\|_k\|v_1\|_d\|v_2\|_d\|w\|_d + \|v_1\|_k\|v_2\|_d\|w\|_d + \|v_1\|_d\|v_2\|_k\|w\|_d + \|v_1\|_d\|v_2\|_d\|w\|_k\} + P_k(\|u\|_{k-1})\|v_1\|_{k-1}\|v_2\|_{k-1}\|w\|_{k-1} .$$

Proof. Recall 4.3.2-3 Lemmas. The local expression of $\widetilde{A}$ is denoted there by Ψ. Thus, using the notations in 4.3.2 - 3 Lemmas, we put $\Phi(u)v = G(0)p\Psi(u)v$. Then, the above conditions (i) and (ii) are satisfied, and the condition (iii) is ensured by 4.3.4 Lemma. The inequalities in (iv) are easily proved by the assumptions.

§ VII Frobenius theorem on strong ILB-Lie groups

6.3.2 Theorem shows that we can apply Frobenius theorem if an involutive distribution is given by the kernel of a right invariant C^∞ILBC^2-normal bundle morphism $\tilde{A}$ and the complementary subspaces E_2^k are Hilbert spaces. However, such mixed (half Banach and half Hilbert) conditions are not satisfied in general. So, we have to restrict our attention to distributions given on strong ILH-Lie groups or finite codimensional distributions on strong ILB-Lie groups. In the last case, we can easily find a finite dimensional (hence Hilbert) complementary subspase.

VII.1 Frobenius theorem on strong ILH-Lie groups.

In this section, we will discuss about a Frobenius theorem on strong ILH-Lie groups by using the results in the previous chapter. (Cf. 6.3.2 Theorem.)

Now, consider a C^∞ILHC^2-normal bundle morphism $\tilde{A}$ of the tangent bundle $B(\mathfrak{g},G,\tilde{T}_\theta)$ into another ILH-vector bundle $B(\mathbb{F},G,\tilde{T}')$ of order $r \geq 0$. Recall that in the definition of $B(\mathfrak{g},G,\tilde{T}_\theta)$ and C^∞ILHC^2-normal bundle morphism, we have to fix an ILH-coordinate (U,ξ) of G at the identity. Assume furthermore that $\tilde{A}$ satisfies the conditions (b) ~ (d) in p.IV.6. The ILH-splitting in the condition (b) will be denoted by $\mathfrak{g} = \mathfrak{h} \oplus \mathbb{E}_2$, $\mathbb{F} = \mathbb{F}_1 \oplus \mathbb{F}_2$. $\mathfrak{h}$ is, then, the kernel of A.

We assume that $\mathfrak{h}$ is a Lie subalgebra of $\mathfrak{g}$. Let $\tilde{\mathfrak{h}} = \{dR_g\mathfrak{h} : g \in G\}$, $\tilde{\mathfrak{h}}^k = \{dR_g\mathfrak{h}^k : g \in G^k\}$. Then, $\tilde{\mathfrak{h}}^k$ is the kernel of $\tilde{A} : B(\mathfrak{g}^k,G^k,\tilde{T}_\theta) \rightarrow B(F^{k-r},G^k,\tilde{T}')$. Therefore, by 4.3.1 Theorem, $\tilde{\mathfrak{h}}^k$ is a smooth subbundle of $B(\mathfrak{g}^k,G^k,\tilde{T}_\theta)$ for every $k \in N(d+r)$. Since $\mathfrak{h}$ is a Lie subalgebra of $\mathfrak{g}$, Proposition A in [31] shows that $\tilde{\mathfrak{h}}^k$ is an involutive distribution.

Let (U,ξ) be the ILH-coordinate of G at e. We translate the above distribution onto that on $U \cap \mathfrak{g}$ through the mapping ξ. Then, 6.3.2 Theorem shows that the resulting distribution on $U \cap \mathfrak{g}$ satisfies all of the conditions of 6.1.3 and 6.2.1. Hence there are open neighborhoods V_1, W_1 of zeros in $\mathfrak{h}^{d+r}$, E_2^{d+r}

respectively and a C^∞ILHC^2-normal mapping Ψ of $V_1 \cap \mathfrak{h} \times W_1 \cap \mathbb{E}_2$ into $U \cap \mathfrak{g}$ with $\Psi(0,0) = 0$, $(d\Psi)_{(o,o)}$ = id., such that $\xi\Psi(V_1 \cap \mathfrak{h}^k, w)$ is an integral submanifold of $\tilde{\mathfrak{h}}^k$ for any $w \in W_1 \cap E_2^k$ and $k \in N(d+r)$.

Put $\xi' = \xi\Psi$. Then, by the inverse function theorem, we may regard $(V_1 \times W_1, \xi')$ as an ILH-coordinate of G at e. This change of coordinates satisfies that $\xi^{-1}\xi'$ is a C^∞ILHC^2-normal mapping with $\xi^{-1}\xi'(0) = 0$ and $(d\xi^{-1}\xi')_o$ = id., and by this mapping ξ', we have slices of integral submanifolds of $\tilde{\mathfrak{h}}$. Remark that $\xi'(V_1 \cap \mathfrak{h}, w)g$ is also an integral submanifold of $\tilde{\mathfrak{h}}$.

Let H be the maximal integral submanifold of $\tilde{\mathfrak{h}}$ through the identity. Since $\tilde{\mathfrak{h}}$ is invariant by right translations, H is a subgroup of G and the restricted mapping $\xi' : V_1 \cap \mathfrak{h} \times \{0\} \mapsto H$ satisfies all conditions (N,1) - (N,7) in § I. Thus, H is a strong ILH-Lie subgroup of G with the Lie algebra $\mathfrak{h}$.

Now, we summarize the results in the following :

7.1.1 Theorem <u>Let</u> G <u>be a strong ILH-Lie group with an ILH-coordinate</u> (U,ξ) <u>at the identity. Let</u> $\tilde{A}$ <u>be a C^∞ILHC^2-normal bundle morphism of the tangent bundle of</u> G <u>into another ILH-vector bundle satisfying (b) ~ (d) in §IV. If a subalgebra</u> $\mathfrak{h}$ <u>of</u> $\mathfrak{g}$ <u>is given by the kernel of</u> $A : \mathfrak{g} \mapsto \mathbb{F}$ (<u>where</u> A <u>is the restriction of</u> $\tilde{A}$ <u>to the fibre at the identity</u>), <u>then there is a strong ILH-Lie subgroup</u> H <u>of</u> G <u>with the Lie algebra</u> $\mathfrak{h}$.

<u>Moreover, there is an ILH-coordinate</u> $(V_1 \times W_1, \xi')$ <u>such that</u> $\xi^{-1}\xi'$ <u>is a C^∞ILHC^2-normal mapping with</u> $\xi^{-1}\xi'(0,0) = 0$, $(d\xi^{-1}\xi')_{(o,o)}$ = id, <u>and that</u> $\xi'(V_1 \cap \mathfrak{h}, w)$ <u>is an integral submanifold of</u> $\tilde{\mathfrak{h}} = \mathrm{Ker}\tilde{A}$ <u>for any</u> $w \in W_1 \cap \mathbb{E}_2$, <u>where</u> $\mathfrak{g} = \mathfrak{h} \oplus E_2$ <u>is an ILH-normal splitting and</u> V_1, W_1 <u>are open neighborhoods of zeros in</u> $\mathfrak{h}^{d+r}$, E_2^{d+r} <u>respectively.</u>

Let $\xi : U \cap \Gamma(T_M) \rightarrow \mathcal{D}$ be an ILH-coordinate mapping defined by $\xi(u)(x) = \mathrm{Exp}\, u(x)$, using a smooth connection on M. Now, let E, F be smooth finite dimensional riemannian vector bundle over M and consider differential operators $A : \Gamma(T_M) \rightarrow \Gamma(E)$, $B : \Gamma(E) \mapsto \Gamma(F)$ of order r with smooth coefficients such that

$BA = 0$ and $AA^* + B^*B$ is an elliptic differential operator, where A^*, B^* are formal adjoint operators of A, B respectively. Assume $\mathfrak{g} = \operatorname{Ker} A$ is a Lie subalgebra of $\Gamma(T_M)$. Then, by 5.2.2 Theorem, we have the following :

7.1.2 Corollary <u>There is a strong ILH-Lie subgroup</u> H <u>of the strong ILH-Lie group</u> $\mathcal{D}$ <u>with the Lie algebra</u> $\mathfrak{g}$. <u>Moreover, there is an ILH-coordinate</u> $(V_1 \times W_1, \xi')$ <u>such that</u> $\xi^{-1}\xi'$ <u>is a</u> C^∞<u>ILHC</u>^{2}<u>-normal mapping with</u> $\xi^{-1}\xi'(0,0) = 0$, $(d\xi^{-1}\xi')_{(0,0)} = \mathrm{id}$. <u>and that</u> $\xi'(V_1 \cap \mathfrak{g}, w)$ <u>is an integral submanifold of</u> $\tilde{\mathfrak{g}} = \{dR_g\mathfrak{g} : g \in \mathcal{D}\}$ <u>for any</u> $w \in W_1 \cap \mathbb{E}_2$.

Let $\xi' : U \cap (\Gamma(1_M) \oplus \Gamma(T_M)) \mapsto \Gamma_*(1_M) * \mathcal{D}$ be the ILH-coordinate mapping defined by $\xi'(f,u) = (e^f, \xi(u))$. Let $A : \Gamma(1_M) \oplus \Gamma(T_M) \mapsto \Gamma(E)$, $B : \Gamma(E) \mapsto \Gamma(F)$ be differential operators of order r with smooth coefficients such that $BA = 0$, and $AA^* + B^*B$ is elliptic. (Since $\Gamma(1_M) \oplus \Gamma(T_M) = \Gamma(1_M \oplus T_M)$, we can define A^* by the same manner as above.) Assume furthermore that $\mathfrak{g} = \operatorname{Ker} A$ is a Lie subalgebra of $\Gamma(1_M) \oplus \Gamma(T_M)$. Then, 5.2.3 Theorem yields the following :

7.1.3 Corollary <u>There is a strong ILH-Lie subgroup</u> H <u>of the strong ILH-Lie group</u> $\Gamma_*(1_M) * \mathcal{D}$ <u>with the Lie algebra</u> $\mathfrak{g}$. <u>Moreover, there is an ILH-coordinate</u> $(V_1 \times W_1, \xi'')$ <u>such that</u> $\xi'^{-1}\xi''$ <u>is a</u> C^∞<u>ILHC</u>^{2}<u>-normal mapping with</u> $\xi'^{-1}\xi''(0,0) = 0$, $(d\xi'^{-1}\xi'')_{(0,0)} = \mathrm{id}$. <u>and that</u> $\xi''(V_1 \cap \mathfrak{g}, w)$ <u>is an integral submanifold of</u> $\tilde{\mathfrak{g}} = \{dR_g\mathfrak{g} : g \in \Gamma_*(1_M) * \mathcal{D}\}$ <u>for any</u> $w \in W_1 \cap \mathbb{E}_2$.

Now, let H be the resulting strong ILH-Lie subgroup of $\mathcal{D}$ (resp. $\Gamma_*(1_M) * \mathcal{D}$). We consider a compact subgroup K of H (resp. $H \cap (\{1\} * \mathcal{D})$). By 1.4.2, K is a Lie group. Recall 3.2.2 ~ 3 Theorems (resp. 3.2.2 ~ 3 and 4.5.4). Then, we see easily the following :

7.1.4 Corollary <u>Notations and assumptions are as in 7.1.2 ~ 3. Let</u> K <u>be a compact subgroup of</u> H <u>or</u> $H \cap (\{1\} * \mathcal{D})$. <u>Then,</u> $H_K = \{h \in H : hk = kh \text{ for any } k \in K\}$ <u>is a strong ILH-Lie subgroup of</u> H.

Now, start with a subgroup H of a strong ILH-Lie group G. Let $\mathfrak{h}$ be the set of all infinitesmal generators u of one parameter subgroups exp tu such that exp tu ∈ H for any t. We assume that $\mathfrak{h}$ is a closed Lie subalgebra of $\mathfrak{g}$ or we assume the same condition as in 1.4.1. Assume furthermore that $\mathfrak{h}$ is a kernel of $A : \mathfrak{g} \mapsto \mathbb{E}$ such that $\tilde{A} = dR_g A dR_{g^{-1}}$ is a C^∞ILHC^2-normal bundle morphism with the conditions (b) ~ (d) of the tangent bundle of G into another ILH-vector bundle $B(\mathbb{E}, G, \tilde{T}')$. Then, by 7.1.1, there is a strong ILH-Lie subgroup H' of G with the Lie algebra $\mathfrak{h}$.

Recall that H' is obtained by Frobenius theorem. So, if a piecewise C^1-curve c(t) in G satisfies c(0) = e and c(t) ∈ H for all t, then we see that $\dot{c}(t) \in \tilde{\mathfrak{h}}$ (cf. 1.4.4.), and hence c(t) ∈ H' for any t. Therefore, we see

7.1.5 Theorem H <u>is a strong ILH-Lie group under the LPSAC-topology.</u> H' <u>is the connected component of</u> H <u>under the LPSAC-topology.</u>

Suppose H is a closed subgroup of G. By 1.4.1 Theorem, $\mathfrak{h}$ is a closed Lie subalgebra of $\mathfrak{g}$. Assume the same condition for $\mathfrak{h}$ as above. Thus, H is a strong ILH-Lie group under the LPSAC-topology. However, in general, the LPSAC-topology for H is stronger than the relative topology in G. Even if H is closed in G, H' may or may not be closed in G. (There is one dimensional closed subgroup H such that H ≠ H'.) The following are sufficient conditions for the above two topologies to coincide:

(1) H is LPSAC. (i.e. Locally piecewise-smooth-arcwise connected.) (Trivial.)

(2) H satisfies the second countability axiom under the LPSAC-topology and G satisfies the second countability axiom.

(3) G satisgies the second countability axiom and H is PSAC. (i.e. Piecewise-smooth-arcwise connected.)

We will explain (2) and (3).

(2) : Recall that $\mathfrak{h}$ satisfies the conditions of 7.1.1 theorem. Thus, there is an

ILH-coordinate $(V_1 \times W_1, \xi')$ of G at e such that $\xi'(V_1 \cap \mathfrak{h}, w)$ is an integral submanifold of $\tilde{\mathfrak{h}}$ for any $w \in W_1 \cap E_2$. Thus, for any closed, star shaped neighborhood V' of 0 in $\mathfrak{h}$ such that $V' \subset V_1 \cap \mathfrak{h}$, we see that $\xi'(V', 0)$ is a closed subset of G and a closed neighborhood of e under the LPSAC-topology. Since G has the second countability axiom, G is a set of second category. Therefore, we can apply the Baire category theorem, and obtain that $\xi'(V',0)$ is a neighborhood of e of H under the relative topology.

(3) : Since G has the second countability axiom, the Lie algebra $\mathfrak{g}$ has the second countability axiom, hence so does $\mathfrak{h}$ as a closed subset of $\mathfrak{g}$. H is a strong ILH-Lie group under the LPSAC-topology and connected. So, H satisfies the second countability axiom because H is generated by an open neighborhood of the identity, which is homeomorphic to an open neighborhood of 0 of $\mathfrak{h}$. Thus, we get the case (2).

VII.2 Finite codimensional subalgebras of strong ILB-Lie groups.

Here we will apply 6.3.2 Theorem to a subalgebra of finite codimension.

Let G be a strong ILB-Lie group with the Lie algebra $\mathfrak{g}$. Let $\mathfrak{h}$ be a closed subalgebra of finite codimension in $\mathfrak{g}$. Let $\mathfrak{h}^k$ be the closure of $\mathfrak{h}$ in $\mathfrak{g}^k$. We assume that $\dim \mathfrak{g}/\mathfrak{h} = \dim \mathfrak{g}^d/\mathfrak{h}^d$. In general, this is not true. We have only $\dim \mathfrak{g}^d/\mathfrak{h}^d \leqslant \dim \mathfrak{g}^k/\mathfrak{h}^k \leqslant \dim \mathfrak{g}/\mathfrak{h}$. Thus, there is $d' \geqslant d$ such that $\dim \mathfrak{g}^{d'}/\mathfrak{h}^{d'} = \dim \mathfrak{g}^k/\mathfrak{h}^k = \dim \mathfrak{g}/\mathfrak{h}$ for any $k \in N(d')$. Therefore, the above assumption is not essential. We may replace d by d'.

Now, by the assumption, there is a finite dimensional subspace $\mathfrak{m}$ of $\mathfrak{g}$ such that $\mathfrak{g} = \mathfrak{h} \oplus \mathfrak{m}$, $\mathfrak{g}^k = \mathfrak{h}^k \oplus \mathfrak{m}$, $k \in N(d)$. We put $\tilde{\mathfrak{h}}^k = \{dR_g \mathfrak{h}^k : g \in G^k\}$ and $\tilde{\mathfrak{m}} = \{dR_g \mathfrak{m} : g \in G^k\}$. Then, by the property (G,7) in §I shows that $\tilde{\mathfrak{m}}$ is a trivial bundle over G^k and a smooth subbundle of the tangent bundle T_{G^k} for every $k \geqslant d$.

Let π be the projection of $\mathfrak{g}$ onto $\mathfrak{m}$ inaccordance with the splitting $\mathfrak{g} = \mathfrak{h} \oplus \mathfrak{m}$. Then, π can be extended to the projection of $\mathfrak{g}^k$ onto $\mathfrak{m}$. Define the mapping

$\tilde{\pi} : T_{G^k} \mapsto \tilde{\mathfrak{m}}$ by $dR_g \pi dR_{g^{-1}}$. Then, this is a right invariant mapping and $\mathrm{Ker}\tilde{\pi}$ is given by $\{dR_g \mathfrak{h}^k : g \in G^k\}$. In general, this is only a continuous mapping. However, since $\dim \mathfrak{m} < \infty$, we see the following :

7.2.1 Lemma <u>For any</u> $k \in N(d+1)$, <u>the mapping</u> $\tilde{\pi} : T_{G^k} \mapsto \tilde{\mathfrak{m}}$ <u>is a C^{k-d-1}-bundle morphism.</u>

Proof. It is enough to show that $\tilde{\pi} : T_{G^k} \mapsto \tilde{\mathfrak{m}}$ is a C^{k-d}-mapping. (Cf.1.3.4.) Since $\tilde{\mathfrak{m}}$ is a trivial bundle over G^k and the trivialization is given by $dR_g v$, we have only to show that the mapping $\mu : T_{G^k} \mapsto \mathfrak{m}$ defined by $\mu(u) = \pi dR_{g^{-1}} u$ is a C^{k-d}-mapping, where g is the base point of $u \in T_{G^k}$.

Let $i : \mathfrak{g}^k \mapsto \mathfrak{g}^d$, $j : G^k \mapsto G^d$ be inclusions. Then we have

$$\pi dR_{g^{-1}} u = \pi i dR_{g^{-1}} u = \pi i dR_{j(g)^{-1}} u.$$

By the property (N,5), we see that $u \mapsto dR_{g^{-1}} u$ is a C^{k-d}-mapping of T_{G^k} into $\mathfrak{g}^d$. Thus, we have that $\pi dR_{g^{-1}} u$ is a C^{k-d}-mapping.

7.2.2 Corollary $\tilde{\mathfrak{h}}^k$ <u>is a C^{k-d-1}-subbundle of</u> T_{G^k} .

Proof. This is because $\tilde{\mathfrak{h}}^k$ is the kernel of $\tilde{\pi} : T_{G^k} \mapsto \tilde{\mathfrak{m}}$ and this is surjective.

For simplicity, we call $\tilde{\mathfrak{h}}$ a C^ρ-distribution on G, where ρ is the function $\rho(k) = k-d-1$.

7.2.3 Definition (1) A pair (U', ξ') is called a <u>C^ρ-ILB-coordinate of</u> G at e if $\xi' : U' \cap \mathfrak{g} \mapsto G$ can be extended to a $C^{\rho(k)}$-diffeomorphism of $U' \cap \mathfrak{g}^k$ onto an open neighborhood $\xi'(U') \cap G^k$ of e for every $k \in N(d+2)$.

(2) A subgroup H of G is called a <u>C^ρ-strong ILB-Lie subgroup of</u> G, if the conditions (Sub. 1 - 2) (cf. p.I.14) are satisfied after replaceing "ILB-coordinate" by "C^ρ-ILB-coordinate".

(3) A right invariant fibre preserving mapping $\Phi : B(\mathbb{E}, G, \tilde{T}) \mapsto B(\mathbb{F}, G, \tilde{T}')$ is called a <u>C^ρILBC^2-normal bundle morphism of order</u> r, if Φ satisfies the inequalities (e) of p.IV.6 replacing d by $d' = d+2$, and the following (a') :

(a') Φ can be extended to a $C^{\rho(k)}$-bundle morphism of $B(E^{k'},G^k,\tilde{T})$ into $B(F^{k'-r},G^k,\tilde{T}')$ for any $k \geq k' \geq d+r+2$.

Remark In the proofs of the theorems in §VI, we do not use the smoothness of the mappings so strictry. It was enough to be 2-times differentiable. So, all the theorems in §VI is still true by replacing C^∞ILB··· by C^ρILB···.

Moreover, Proposition A in [31] is also true for C^ρ-distributions. Namely, if $\tilde{\mathfrak{h}}^k$ is a C^r-distribution on G^k which is right invariant and $\mathfrak{h}$ is a Lie subalgebra, then $\tilde{\mathfrak{h}}^k$ is a C^r- involutive distribution on G^k for $r \geq 1$.

7.2.4 Theorem $\tilde{\pi} : T_G \longmapsto \tilde{\mathfrak{m}}$ is a C^ρILBC^2-normal bundle morphism of order 0 satisfying the conditions (b) ~ (d) in p.IV.6. Thus, 6.3.2 Theorem and all the theorems in §VI.1 - 2 can be applied, hence there is a C^ρ-strong ILB-Lie subgroup H of G with the Lie algebra $\mathfrak{h}$.

Proof. We have only to check the conditions (a') and (b) ~ (e) in p.IV.6. However we have to change the norms as in 3.4.1. Since $\mathfrak{h}$ is of finite codimension, we can define the new norm $[\![\ ]\!]_k$ by the same manner as in 3.4.1.

Now, put $r = 0$ in the conditions (a'), (b) ~ (e). Then, (a') and (b) are already checked and (c) is trivial in this case. (d) is easy to see.

Let (U,ξ) be an ILB-coordinate of G at e. The local expression of $\tilde{\pi}$ is given by $\Phi(u)w = dR^{-1}_{\xi(u)}\, \pi\, (d\xi)_u w$. The mapping $\Phi : U \cap \mathfrak{g} \times \mathfrak{g} \longmapsto \mathfrak{m}$ can be extended to the C^{k-d}-mapping of $U \cap \mathfrak{g}^k \times \mathfrak{g}^k$ onto $\mathfrak{m}$. Especially, putting $d' = d+2$, we have that Φ can be extended to a C^2-mapping of $U \cap \mathfrak{g}^{d'} \times \mathfrak{g}^{d'}$ onto $\mathfrak{m}$. Thus, there is an open star shaped neighborhood W of 0 in $\mathfrak{g}^{d'}$ such that

$$\|(d_1\Phi)_u(v,w)\|_{d'} \leq C\|v\|_{d'}\|w\|_{d'}\ , \quad \|(d_1^2\Phi)_u(v_1,v_2,w)\|_{d'} \leq C\|v_1\|_{d'}\|v_2\|_{d'}\|w\|_{d'}\ .$$

Thus, $\|\Phi(u)w\|_{d'} - \|\Phi(0)w\|_{d'} \leq C\|u\|_{d'}\|w\|_{d'}$, and hence

$$\|\Phi(u)w\|_{d'} \leq C\{\|u\|_{d'}\|w\|_{d'} + \|w\|_{d'}\},$$

because $\Phi(0) = id.$.

Remark that $[\![\Phi(u)w]\!]_k = \|\Phi(u)w\|$ etc.. The desired inequalities follow from the above inequalities and the fact $\|u\|_k \geq \|u\|_{d'}$ for any $u \in \mathfrak{g}^k$, $k \in N(d')$.

7.2.5 Corollary Notations and assumptions being as above, if moreover H is closed in G, then H^k is closed in G^k, $k \geq d+2$ and the factor set $H^k\backslash G^k$ is a finite dimensional smooth manifold. Moreover, $H^k\backslash G^k = H\backslash G$ and every element of G^k acts on $H\backslash G$ as $C^{\rho(k)}$-diffeomorphism.

Proof. Assume H^k is not closed in G^k. Then, there is a sequence $\{w_n\}$ in $\mathfrak{m}$ such that $w_n \neq 0$, $\{w_n\}$ converges to 0 in $\mathfrak{h}^k \oplus \mathfrak{m}$ and that $\xi\Psi(0,w_n) \in H^k$, where Ψ is the coordinate mapping naturally obtained in Frobenius theorem. (Cf. §VI.1.) Since $w_n \in \mathfrak{m}$, the convergence of $\{w_n\}$ in $\mathfrak{h}^k \oplus \mathfrak{m}$ implies also the convergence in $\mathfrak{h} \oplus \mathfrak{m}$. Since $w_n \in \mathfrak{g}$, we see that $\Psi(0,w_n) \in H$. Obviously, this is a contradiction because H is closed in G.

If we take a sufficiently small neighborhood W_2 of 0 of $\mathfrak{m}$, then $\xi\Psi(0,W_2)$ gives a local section of $\{H^k g : g \in G^k\}$ for any $k \in N(d+2)$. Therefore, $H^k\backslash G^k$ is a finite dimensional C^{k-d-1}-manifold, on which G^k acts as C^{k-d-1}-diffeomorphisms, where the action A'_g is given by $A'_g(H^k g') = H^k g'g$, hence we have $A'_{g'g} = A'_g A_{g'}$. Similarly, we have that $N = H\backslash G$ is a finite dimensional smooth manifold.

It remains to prove that $N = H^k\backslash G^k$. It is easy to see that N is an open subset of $H^k\backslash G^k$. Since G is dense in G^k, N is dense in $H^k\backslash G^k$. For any point $H^k g$ in $H^k\backslash G^k$, there is $g' \in G$ such that $gg'^{-1} \in \xi\Psi(W_1 \cap \mathfrak{h}^k, W_2) \subset H^k\cdot\xi\Psi(0,W_2)$. Since $\xi\Psi(0,W_2) \subset G$, we see that there is $g'' \in G$ such that $H^k g = H^k g''$. Thus, $N = H^k\backslash G^k$.

VII.3 Left invariant finite codimensional distributions.

Let G be a strong ILB-Lie group modeled on a Sobolev chain $\{\mathfrak{g}, \mathfrak{g}^k, k \in N(d)\}$. $\mathfrak{g}$ is identified with its Lie algebra. Let $\{\mathbb{F}, \mathbb{F}^k, k \in N(d)\}$ another Sobolev chain. Denote by $GL(\mathbb{F})$ the group of all invertible element of $L(\mathbb{F},\mathbb{F})$ (the space of all continuous linear mappings). An anti-homomorphism φ of G into $GL(\mathbb{F})$ is called an ILB-representation of G, if the mapping $\Phi : G \times \mathbb{F} \mapsto \mathbb{F}$, $\Phi(g,f) = \varphi(g)f$, can be extended to a C^ℓ-mapping of $G^k \times \mathbb{F}^{k+\ell}$ into $\mathbb{F}^k$.

Let F^{*k} (resp. $\mathbb{F}^*$) be the dual space of F^k (resp. $\mathbb{F}$). The dual mapping $\varphi^*(g)$ of $\varphi(g)$ is an element of $L(F^{*k}, F^{*k+\ell})$ for every $g \in G^k$. Let Φ^* be the mapping defined by $\Phi^*(g,f') = \varphi^*(g)f'$. Then, we have

7.3.1 Lemma $\varphi^* : G^k \to L(F^{*k}, F^{*k+\ell})$ <u>is a $C^{\ell-1}$-mapping, hence so is</u> $\Phi^* : G^k \times F^{*k} \mapsto F^{*k+\ell}$.

Proof. Consider the mapping $* : L(F^{k+\ell}, F^k) \mapsto L(F^{*k}, F^{*k+\ell})$, $A \to A^*$. Then, this is a continuous linear operator. As $\varphi : G^k \mapsto L(F^{k+\ell}, F^k)$ is a $C^{\ell-1}$-mapping (cf.1.3.4) $\varphi^* : G^k \to L(F^{*k}, F^{*k+\ell})$ is also a $C^{\ell-1}$-mapping.

Now, suppose there is an element $\delta \in F^{*d}$ such that $\dim(d_1\Phi^*)_{(e,\delta)}\mathfrak{g}^d$ is finite. We call such an element a <u>dirac element</u> of F^{*d}. Since $\mathfrak{g}$ is dense in $\mathfrak{g}^k$, we see $(d_1\Phi^*)_{(e,\delta)}\mathfrak{g} = (d_1\Phi^*)_{(e,\delta)}\mathfrak{g}^k$ for any $k \geq d$. Let $\Phi^*_\delta : G^d \mapsto F^{*d+\ell}$ be the $C^{\ell-1}$-mapping defined by $\Phi^*_\delta(g) = \Phi^*(g,\delta) = \varphi^*(g)\delta$. Since the inclusion $G^k \subset G^d$ is smooth, we see that $\Phi^*_\delta : G^k \mapsto F^{*d+\ell}$ is a $C^{\ell-1}$-mapping for every $\ell \geq 1$.

7.3.2 Lemma <u>The kernel of</u> $d\Phi^*_\delta : T_{G^k} \mapsto F^{*d+2}$ <u>is a left invariant smooth distribution of</u> G^k <u>for every</u> $k \in N(d)$, <u>where "left invariant" means that</u> $dL_g \mathrm{Ker} d\Phi^*_\delta = \mathrm{Ker} d\Phi^*_\delta$ <u>for every</u> $g \in G^{k+1}$. (Recall that the left translation L_g is C^1 if $g \in G^{k+1}$.)

Proof. Since $\Phi^*_\delta(L_g h) = \varphi^*(g)\Phi^*_\delta(h)$, if $g \in G^{k+1}$, then $(d\Phi^*_\delta)_{gh}(dL_g)_h = \varphi^*(g)(d\Phi^*_\delta)_h$, hence we have the left invariant property of the kernel.

Let $\xi : U \cap \mathfrak{g} \mapsto G$ be an ILB-coordinate mapping. Let $\mathfrak{m}$ denote the finite dimensional subspace of F^{*d+2} given by $\mathfrak{m} = d(\Phi^*_\delta\xi)_0\mathfrak{g}^d$ ($= d(\Phi^*_\delta\xi)_0\mathfrak{g}$). Denote by $\mathfrak{n}^{*d+2}$ the complementary space of $\mathfrak{m}$ in F^{*d+2}. As $F^{k+\ell} \subset F^k$, we see $F^{*k} \subset F^{*k+\ell}$. Thus, take the closure of $\mathfrak{n}^{*d+2}$ in $F^{*d+\ell}$, $\ell \geq 2$, and denote this by $\mathfrak{n}^{*d+\ell}$. We have easily that $F^{*d+\ell} = \mathfrak{m} \oplus \mathfrak{n}^{*d+\ell}$ for every $\ell \geq 2$. Let π be the projection of F^{*d+2} onto $\mathfrak{m}$ inaccordance with the above splitting. Then, π can be extended to the projection of $F^{*d+\ell}$ onto $\mathfrak{m}$, hence the projection of $\mathbb{F}^*$ onto $\mathfrak{m}$.

There exists a finite dimensional subspace $\mathfrak{m}'$ of $\mathfrak{g}$ such that $d(\Phi^*_\delta\xi)_0 : \mathfrak{m}' \mapsto \mathfrak{m}$ is an isomorphism. Thus, there is an open neighborhood W of 0 in $\mathfrak{g}^d$

such that $\pi(d\Phi_\delta^*\xi)_u : \mathfrak{m}' \longrightarrow \mathfrak{m}$ is an isomorphism for every $u \in W \cap \mathfrak{g}^k$, $k \in N(d)$.

Since the mapping $\pi d(\Phi_\delta^*\xi) : W \cap \mathfrak{g}^k \times \mathfrak{g}^k \longmapsto \mathfrak{m}$ can be regarded as a mapping into $F^{*d+\ell}$ for every $\ell \geq 2$, the above mapping is a smooth mapping. Since this is surjective the kernel of $\pi d(\Phi_\delta^*\xi)$ defines a smooth distribution on $W \cap \mathfrak{g}^k$ for every $k \in N(d)$.

So, to complete the proof, it is enough to show that $\mathrm{Ker}\pi d(\Phi_\delta^*\xi) = \mathrm{Ker} d(\Phi_\delta^*\xi)$. Since $\pi : d(\Phi_\delta^*\xi)_u \mathfrak{m}' \longmapsto \mathfrak{m}$ is an isomorphism for every $u \in W \cap \mathfrak{g}^k$, if $\pi d(\Phi_\delta^*\xi)_u v = 0$ for $v \in \mathfrak{g}^k$, then we have $d(\Phi_\delta^*\xi)_u v = 0$, hence $\mathrm{Ker}\pi d(\Phi_\delta^*\xi)_u = \mathrm{Ker} d(\Phi_\delta^*\xi)_u$ for every $u \in W \cap \mathfrak{g}^k$.

Now, let $\mathfrak{m} = (d\Phi_\delta^*)_e \mathfrak{g}^d = (d\Phi_\delta^*)_e \mathfrak{g}$ and $\mathfrak{h} = \mathrm{Ker}\{(d\Phi_\delta^*)_e : \mathfrak{g} \longrightarrow F^{*d+2}\}$. Then, ovbiously $\mathfrak{h}$ is a finite ($\dim \mathfrak{m}$) codimensional subalgebra of $\mathfrak{g}$. Let $\mathfrak{h}^k$ be the closure of $\mathfrak{h}$ in $\mathfrak{g}^k$. Then, $\mathfrak{h}^k$ is the kernel of $(d\Phi_\delta^*)_e : \mathfrak{g}^k \mapsto F^{*k+2}$. There is a finite dimensional subspace $\mathfrak{m}'$ of $\mathfrak{g}$ such that $\mathfrak{g} = \mathfrak{h} \oplus \mathfrak{m}'$, $\mathfrak{g}^k = \mathfrak{h}^k \oplus \mathfrak{m}'$, $k \in N(d)$.

Let $\widetilde{\ell\mathfrak{h}} = \{dL_g \mathfrak{h} : g \in G\}$. Then, we see easily that $\widetilde{\ell\mathfrak{h}}$ is the kernel of $d\Phi_\delta^* : T_G \mapsto F^{*d+2}$. Hence, by the above lemma, we see that $\widetilde{\ell\mathfrak{h}}$ can be extended to a smooth distribution $\widetilde{\ell\mathfrak{h}}^k = \mathrm{Ker}\{ d\Phi_\delta^* : T_{G^k} \longmapsto F^{*d+2} \}$ on G^k for every $k \in N(d)$.

7.3.3 Lemma $\widetilde{\ell\mathfrak{h}}^k$ <u>is an involutive distribution for any</u> $k \in N(d)$.

Proof. We have only to show that $\widetilde{\ell\mathfrak{h}}^k$ is involutive on a neighborhood of e. Let $\nu : U \cap \mathfrak{g} \times \mathfrak{g} \mapsto \mathfrak{m}'$ be the mapping defined by $\nu(u)v = j\pi d(\Phi_\delta^*\xi)_u v$, where j is the inverse of $(d\Phi_\delta^*\xi)_o : \mathfrak{m}' \longmapsto \mathfrak{m}$. Since $\nu(0) : \mathfrak{m}' \longmapsto \mathfrak{m}'$ is the identity, there is an open neighborhood W of 0 of $\mathfrak{g}^d$ such that $\nu(u) : \mathfrak{m}' \mapsto \mathfrak{m}'$ is an isomorphism for every $u \in W$. Let $G(u)$ be its inverse. Then, $G : W \cap \mathfrak{g}^k \rightarrow GL(\mathfrak{m}')$ is a C^∞-mapping for every $k \in N(d)$.

For any $v \in \mathfrak{h}^k$, we see that $v - G(u)\nu(u)v$ is contained in $\mathrm{Ker}\ \nu(u)$. Moreover, $\mathrm{Ker}\ \nu(u) = \{ v - G(u)\nu(u)v : v \in \mathfrak{h}^k\}$.

Let $\Theta : W \cap \mathfrak{g} \times \mathfrak{h} \oplus \mathfrak{m}' \rightarrow \mathfrak{h} \oplus \mathfrak{m}'$ be the mapping defined by $\Theta(u)(h + f) = h - G(u)\nu(u)h + f$. Then, Θ can be extended to a C^∞-mapping of $W \cap \mathfrak{g}^k \times \mathfrak{h}^k \oplus \mathfrak{m}'$ into $\mathfrak{h}^k \oplus \mathfrak{m}'$ and $\Theta : W \cap \mathfrak{g}^k \longmapsto L(\mathfrak{h}^k \oplus \mathfrak{m}', \mathfrak{h}^k \oplus \mathfrak{m}')$ is a C^∞-mapping.

It is easy to see that $\Theta(u)$ is bijective for every $u \in W \cap \mathfrak{G}^k$.

The mapping Θ induces a bundle isomorphism $\tilde{\Theta} : W \cap \mathfrak{G} \times \mathfrak{G} \longrightarrow W \cap \mathfrak{G} \times \mathfrak{G}$, $\tilde{\Theta}(u,v) = (u, \Theta(u)v)$, which can be extended to a C^∞-bundle isomorphism of $W \cap \mathfrak{G}^k \times \mathfrak{G}^k$ onto itself for every $k \in N(d)$. By the definition, we have $\tilde{\Theta}(u, \mathfrak{h}^k) = (u, \mathrm{Ker}\,\nu(u))$.

Now, for every C^1-vector field X on $W \cap \mathfrak{G}$ in the sense of Frechet manifolds, the pull back $\tilde{\Theta}^{-1}X$ is a C^1-section of $W \cap \mathfrak{G} \times \mathfrak{G}$ such that if X is a vector field contained in the distribution $d\xi^{-1}\tilde{\ell\mathfrak{h}}$, then $\tilde{\Theta}^{-1}X$ is a C^1-section of $W \cap \mathfrak{G} \times \mathfrak{h}$.

We define a connection ∇ on $W \cap \mathfrak{G}$ by the following : For C^1-vector fields X, Y on $W \cap \mathfrak{G}$, we put $\nabla_X Y = \tilde{\Theta}(d\tilde{\Theta}^{-1}Y)(\tilde{\Theta}^{-1}X)$, where d in the right hand side means the derivative of $\tilde{\Theta}^{-1}Y$. This connection ∇ can be extended to a C^∞-connection on $W \cap \mathfrak{G}^k$ for any $k \in N(d)$. Moreover, if X, Y are C^1-vector fields contained in $\mathrm{Ker}\,\nu$, then so is $\nabla_X Y$.

Let $T(X,Y) = \nabla_X Y - \nabla_Y X - [X,Y]$. Then, T is called a torsion tensor and can be extended to a C^∞-tensor field on $W \cap \mathfrak{G}^k$.

For any $u, v \in \mathfrak{h}$, we put $\tilde{u}(w) = d\xi^{-1}dL_{\xi(w)}u$, $\tilde{v}(w) = d\xi^{-1}dL_{\xi(w)}v$. These are smooth vector fields on $W \cap \mathfrak{G}$ contained in $d\xi^{-1}\tilde{\ell\mathfrak{h}}$. Since $[\tilde{u}, \tilde{v}] = [u, v]^\sim$, we see that $T(\tilde{u}, \tilde{v}) = \nabla_{\tilde{u}}\tilde{v} - \nabla_{\tilde{v}}\tilde{u} - [u,v]^\sim \in d\xi^{-1}\tilde{\ell\mathfrak{h}}$. Therefore, for every $w \in W \cap \mathfrak{G}$, we have $T_w(d\xi^{-1}\tilde{\ell\mathfrak{h}}, d\xi^{-1}\tilde{\ell\mathfrak{h}}) \subset d\xi^{-1}\tilde{\ell\mathfrak{h}}$, where T_w means the torsion tensor at w. Since T can be extended to a C^∞-tensor field on $W \cap \mathfrak{G}^k$, we see that $T_w(\mathrm{Ker}\,\nu, \mathrm{Ker}\,\nu) \subset \mathrm{Ker}\,\nu$ for every $w \in W \cap \mathfrak{G}^k$. This implies that $\mathrm{Ker}\,\nu$ is involutive.

Now, from this moment, we can forget about that the distribution is left invariant. We may simply start with $\nu : U \cap \mathfrak{G} \times \mathfrak{G} \longrightarrow \mathfrak{m}'$ such that ν can be extended to a C^∞-mapping of $U \cap \mathfrak{G}^k \times \mathfrak{G}^k \longrightarrow \mathfrak{m}'$ and the kernel of ν gives a smooth distribution on $U \cap \mathfrak{G}^k$. Moreover, since $\dim \mathfrak{m}' < \infty$, we see that ν is a C^∞ILBC^2-normal bundle morphism of order 0. (Cf.6.2.1.) Thus, we can apply all of the theorems in §VI. Therefore, we get the Frobenius theorem for the distribution $\tilde{\ell\mathfrak{h}}$.

7.3.4 Theorem <u>Let $\varphi : G \mapsto GL(\mathbb{F})$ be an ILB-representation of</u> G. <u>Suppose we have a dirac element δ in the dual space F^{*d} of F^d. Then,</u> $G_\delta = \{ g \in G : \varphi^*(g)\delta = \delta \}$ <u>is a closed strong ILB-Lie subgroup of</u> G. <u>Moreover, the factor set</u> G/G_δ (left coset) <u>is a smooth finite dimensional manifold such that</u> $G/G_\delta = G^k/G^k_\delta$ <u>and is immersed in</u> $F^{*d+\ell}$ <u>as a $C^{\ell-1}$-submanifold.</u>

Proof. By the above argument, we see that there is a strong ILB-Lie subgroup H of G with the Lie algebra $\mathfrak{h}$. By the same reasoning of 6.1.5, we see that H is a connected component of G_δ under the LPSAC-topology.

Since $\Phi^*_\delta : G^k \mapsto F^{*d+\ell}$ is a $C^{\ell-1}$-mapping and $(d\Phi^*_\delta)_e \mathfrak{g}^k$ is of finite dimension, we can use the implicit function theorem on Banach manifolds. Thus, we see that $G^k_\delta = \{ g \in G^k : \varphi(g)\delta = \delta \}$ is a $C^{\ell-1}$-submanifold of G^k. (Cf. the argument in 3.4.1.) Recall that H^k is obtained by the Frobenius theorem. Thus, by the same reasoning as in 6.1.5, we see that H^k is the connected component of G^k_δ containing the identity.

Therefore, H is the connected component of G_δ, hence G_δ is a strong ILB-Lie subgroup of G. Obviously, G_δ is closed in G.

Since we have a local section of the slices of integral submanifolds, we see that G/G_δ is a C^∞-submanifold and is equal to G^k/G^k_δ by the same proof as in 7.2.5.

$\Phi^*_\delta : G^k \longmapsto F^{*d+\ell}$ is a $C^{\ell-1}$-mapping, hence we see that G^k/G^k_δ is immersed in $F^{*d+\ell}$ as a $C^{\ell-1}$-submanifold.

<u>Remark</u> Using only the implicit function theorem on Banach manifolds, we can conclude that G_δ is an ILB-Lie group of G. However, this method does not ensure that G_δ is a <u>strong</u> ILB-Lie subgroup of G.

<u>Note.</u> The author does not know the structure of the set of all dirac elements in F^{*d}. It is very likely to be a stratified set.

§ VIII Miscellaneous examples

In this chapter, we will give miscellaneous examples of strong ILH-Lie groups and the proofs of Theorems I,J,K in § 0.

Recall that we have already three concrete examples of strong ILH-Lie groups as follows :

(1) $\mathcal{D}$: The group of all C^∞-diffeomorphisms on a smooth, closed manifold M.

(2) $\mathcal{D}_{\mathcal{F}}$: Letting $\mathcal{F}$ be a fibering of M with a compact fibre, the group of all fibre preserving diffeomorphisms.

(3) $\Gamma_*(1_M)*\mathcal{D}$: Semi-direct product of $\Gamma_*(1_M)$ and $\mathcal{D}$.

As a matter of course, if we use the implicit function theorem of §III, then we can get other examples. Moreover, every finite or infinite dimensional Hilbert Lie groups are strong ILH-Lie groups.

VIII.1 Volume preserving transformations.

Here, M is oriented, closed, riemannian manifold. Let dV be a volume element on M defined by the riemannian metric on M. We put $\mathcal{D}_{dV} = \{ \varphi \in \mathcal{D} : \varphi^* dV = dV\}$. The goal here is to prove the following :

8.1.1 Theorem $\mathcal{D}_{dV}$ <u>is a strong ILH-Lie subgroup of</u> $\mathcal{D}$.

Proof. We consider the space $\Gamma_{dV}(T_M) = \{ u \in \Gamma(T_M) : \operatorname{div} u = 0 \}$. $\Gamma_{dV}(T_M)$ is the space of all infinitesimal volume preserving transformations and then, forms a Lie subalgebra of $\Gamma(T_M)$.

Consider differential operators $A = \operatorname{div} : \Gamma(T_M) \mapsto \Gamma(1_M)$, $B : \Gamma(1_M) \mapsto \{0\}$ of order 1. Since $(\operatorname{div})^* = \operatorname{grad}$, we see that div grad is elliptic. Thus, 4.3.1, 5.2.2 and 7.1.1 can be applied and hence we have a strong ILH-Lie subgroup $\mathcal{D}'_{dV}$ of $\mathcal{D}$ with the Lie algebra $\Gamma_{dV}(T_M)$, where $\mathcal{D}^k_{dV}$ is defined for $k \geq \dim M + 7$.

Let $\mathcal{D}^k_{dV} = \{ \varphi \in \mathcal{D}^k : \varphi^* dV = dV\}$. Remark that we can not conclude directly that

$\mathcal{D}'_{dV} = \mathcal{D}_{dV}$. However, we see easily that $\mathcal{D}'_{dV} \subset \mathcal{D}_{dV}$ and $\mathcal{D}'^{k}_{dV} \subset \mathcal{D}^{k}_{dV}$. Moreover, since $\mathcal{D}'_{dV}$ is obtained by the Frobenius theorem, if a piecewise C^1-curve $c(t)$ satisfies $c(0) = e$, $c(t) \in \mathcal{D}^{k}_{dV}$, then $c(t) \in \mathcal{D}'^{k}_{dV}$. Thus, to prove $\mathcal{D}'^{k}_{dV} = \mathcal{D}^{k}_{dV,o}$ (the connected component of $\mathcal{D}^{k}_{dV}$ containing the identity,) we have only to show that $\mathcal{D}^{k}_{dV}$ is LPSAC. However, it is known by Ebin and Marsden [10] that $\mathcal{D}^{k}_{dV}$ is a smooth Hilbert submanifold of $\mathcal{D}^{k}$. They proved this by using the implicit function theorem in Hilbert manifolds. (See also VIII.3 in this chapter, where we will give the precise proof for the contact transformation groups by the almost parallel manner.) Thus, we have $\mathcal{D}'^{k}_{dV} = \mathcal{D}^{k}_{dV,o}$ and hence $\mathcal{D}'_{dV} = \mathcal{D}_{dV,o}$. Therefore, $\mathcal{D}_{dV}$ is a closed, strong ILH-Lie subgroup of $\mathcal{D}$.

Remark 1 By the same method of 8.7.5 Lemma of this chapter, we know that the above result can be given by using the implicit function theorem of § III.

Remark 2 In the above argument, we have to assume that M is oriented. If M is not an orientable manifold, then we consider double covering $\tilde{M}$ of M. Let Z_2 be the deck transformation group. $\mathcal{D}(M)$ is naturally identified with $\mathcal{D}_{Z_2}(\tilde{M})$. (Cf. 2.2.1.) Obviously, Z_2 is contained in $\mathcal{D}_{dV}(\tilde{M})$. Thus, by 3.2.3 and 7.1.2, we see that $\mathcal{D}_{dV,Z_2}(\tilde{M}) = \{ \varphi \in \mathcal{D}_{dV}(\tilde{M}) : \varphi k = k\varphi , k \in Z_2 \}$ is a strong ILH-Lie subgroup of $\mathcal{D}_{dV}(\tilde{M})$. $\mathcal{D}_{dV,Z_2}(\tilde{M})$ can be naturally identified with the volume preserving transformation group on M, i.e. $|\varphi^* dV| = |dV|$.

VIII.2 Symplectic transformations.

Here, M is oriented and even dimensional, say 2m-dimensional. Assume there is a smooth symplectic 2-form Ω, i.e. closed 2-form Ω with the property $\Omega^m \neq 0$. Let T^*_M be the cotangent bundle of M. Then, $\Omega \lrcorner : T_M \mapsto T^*_M$ is a bundle isomorphism, where $\Omega \lrcorner X$ means the inner product or "insert".

The Lie algebra of infinitesimal symplectic transformations is given by

$$\Gamma_\Omega(T_M) = \{ u \in \Gamma(T_M) : \mathcal{L}_u\Omega = d(\Omega \lrcorner u) = 0 \},$$

where $\mathcal{L}$ means the Lie derivative.

Let Λ_M^i be the i-th exterior product of T_M^*. Define differential operators A and B by $A = d(\Omega \lrcorner *) : \Gamma(T_M) \mapsto \Gamma(\Lambda_M^2)$ and $B = d : \Gamma(\Lambda_M^2) \mapsto \Gamma(\Lambda_M^3)$. It is easy to see that $AA^* + B^*B$ is elliptic because $d\delta + \delta d$ is elliptic, where $\delta = d^*$. Obviously we have $BA = 0$. Thus, by 4.3.1, 5.2.2 and 7.1.1, there is a strong ILH-Lie subgroup $\mathcal{D}'_\Omega$ of $\mathcal{D}$ with the Lie algebra $\Gamma_\Omega(T_M)$, where $\mathcal{D}'^k_\Omega$ is defined for $k \geq \dim M + 7$.

Let $\mathcal{D}_\Omega = \{ \varphi \in \mathcal{D} : \varphi^*\Omega = \Omega \}$, $\mathcal{D}^k_\Omega = \{ \varphi \in \mathcal{D}^k : \varphi^*\Omega = \Omega \}$. We can not conclude directly that $\mathcal{D}'_\Omega = \mathcal{D}_{\Omega,o}$ (the connected component). To prove this, we have to know that $\mathcal{D}^k_\Omega$ is LPSAC. However, Ebin and Marsden [10] show that $\mathcal{D}^k_\Omega$ is a smooth Hilbert manifold of $\mathcal{D}^k$. (See also VIII.8 of this chapter, where we have a stronger result.) Thus, we can get $\mathcal{D}'^k_\Omega = \mathcal{D}^k_{\Omega,o}$ and $\mathcal{D}'_\Omega = \mathcal{D}_{\Omega,o}$. Therefore, we get the following :

8.2.1 Theorem $\mathcal{D}_\Omega$ <u>is a closed, strong ILH-Lie subgroup of</u> $\mathcal{D}$.

<u>Remark</u> By the same method of 8.7.4 Lemma of this chapter, we can get the same result by using the implicit function theorem of §III.

VIII.3 Contact transformations.

Here, M is oriented and odd (say 2m+1) dimensional. Assume there is a smooth contact 1-form ω on M, i.e. a 1-form such that $\omega \wedge (d\omega)^m \neq 0$. Although it is usual that the Lie algebra of infinitesimal contact transformations are defined as a subalgebra of $\Gamma(T_M)$ by $\Gamma_\omega(T_M) = \{ u \in \Gamma(T_M) : \mathcal{L}_u\omega = d(\omega \lrcorner u) + d\omega \lrcorner u = f\omega$ for a C^∞-function f $\}$, we consider, here, the subalgebra Γ_ω of $\Gamma(1_M) \oplus \Gamma(T_M)$ (the Lie algebra of $\Gamma_*(1_M)*\mathcal{D}$), which is given by

$$\Gamma_\omega = \{ (f,u) \in \Gamma(1_M) \oplus \Gamma(T_M) : f\omega + d(\omega \lrcorner u) + d\omega \lrcorner u = 0 \},$$

where the Lie algebra structure of $\Gamma(1_M) \oplus \Gamma(T_M)$ is given by 4.5.2 Lemma. Using the general identity $\mathcal{L}_{[u,v]} = \mathcal{L}_v \mathcal{L}_u - \mathcal{L}_u \mathcal{L}_v$, it is easy to see that Γ_ω is a subalgebra of $\Gamma(1_M) \oplus \Gamma(T_M)$.

Now, we consider the differential operators A, B as follows :

$$A : \Gamma(1_M) \oplus \Gamma(T_M) \mapsto \Gamma(T_M^*) \oplus \Gamma(\Lambda_M^2),$$

$$A(f,u) = (f\omega + d(\omega \lrcorner u) + d\omega \lrcorner u,\ d(f\omega) + d(d\omega \lrcorner u)),$$

$$B : \Gamma(T_M^*) \oplus \Gamma(\Lambda_M^2) \mapsto \Gamma(\Lambda_M^2) \oplus \Gamma(\Lambda_M^3), \quad B(\alpha,\beta) = (d\alpha - \beta,\ d\beta).$$

Obviously, we see that $BA = 0$.

8.3.1 Lemma <u>There exists a one to one linear correspondence between</u> Γ_ω <u>and</u> $\Gamma(1_M)$.

Proof. We define a vector field ξ_ω by $\omega \lrcorner \xi_\omega \equiv 1$, $d\omega \lrcorner \xi_\omega \equiv 0$. This is uniquely determined and a smooth vector field on M. Define a subbundle E_ω by $\omega = 0$. Let $\hat{E}_\omega$ be the anihilator of ξ_ω in T_M^*. Then, we have $T_M = R\xi_\omega \oplus E_\omega$, $T_M^* = R\omega \oplus \hat{E}_\omega$. Since $\omega \wedge (d\omega)^m \neq 0$, the mapping $v \mapsto d\omega \lrcorner v$ is an isomorphism of E_ω onto $\hat{E}_\omega$. So, we denote by $d\omega^{-1}$ the inverse mapping.

Now, assume a pair (f,u) satisfies $\mathcal{L}_u\omega + f\omega = 0$. Then, putting $u = h\xi_\omega + v$, $v \in \Gamma(E_\omega)$, we have $\mathcal{L}_u\omega + f\omega = dh + d\omega \lrcorner v + f\omega = 0$. Thus, $-dh\ \xi_\omega = f$. Moreover, we see that $dh - (dh \lrcorner \xi_\omega)\omega$ is contained in $\Gamma(\hat{E}_\omega)$ for any function h. So, the correspondence is given by

$$h \longleftrightarrow (- dh \lrcorner \xi_\omega\ ,\ h\xi_\omega + d\omega^{-1}(dh - (dh \lrcorner \xi_\omega)\omega\ ).$$

We keep the notations as in the proof of the above lemma. Let $E_{\omega,x}$, $\hat{E}_{\omega,x}$ be the fibres of E_ω, $\hat{E}_\omega$ at x respectively. Let $e_1,\dots,e_m,e_{m+1},\dots,e_{2m}$ be a basis of $E_{\omega,x}$. Then, the dual basis of $\xi_\omega(x),e_1,\dots,e_{2m}$ is given by $\omega(x),e_1^*,\dots,e_{2m}^*$, where $e_i^* \in \hat{E}_{\omega,x}$ and $e_i^*(e_j) = \delta_{ij}$. For a suitable choice of $e_1,\dots,e_{2m}$, we may assume that $d\omega \lrcorner e_i = e_{m+i}^*$, $d\omega \lrcorner e_{m+i} = - e_i^*$ for $1 \leq i \leq m$. We call such a basis $\xi_\omega(x)$, $e_1,\dots,$ e_{2m} a symplectic frame at x. All symplectic frames form a smooth principal bundle FSp over M with the fibre Sp(m). However, since the maximal compact subgroup of Sp(m) is U(2m), there is a smooth principal subbundle FU of FSp with the fibre U(2m). This implies that there exists a smooth riemannian metric on M such that $\|\xi_\omega\| = 1$, ξ_ω is perpendicular to E_ω and $< d\omega \lrcorner v, \alpha >_{\hat{E}_\omega}$ = $< v, d\omega^{-1}\alpha >_{E_\omega}$ for every $v \in E_\omega$, $\alpha \in \hat{E}_\omega$.

We denote by δ the formal adjoint operator of the exterior derivative d with respect to the riemannian metric defined above, that is, $< \delta\alpha,\beta >_o = < \alpha, d\beta >_o$, where $< \ , \ >_o$ is given by $< \alpha,\beta >_o = \int_M < \alpha, \beta > dV$.

8.3.2 Lemma <u>The formal adjoint operators of</u> A, B <u>are given by</u>

$$A^*(\alpha,\beta) = (< \omega, \alpha + \delta\beta >_{T^*_M} , \ (\delta\alpha)\xi_\omega + d\omega^{-1}(\alpha + \delta\beta - < \alpha + \delta\beta, \omega >_{T^*_M}\omega))$$

$$B^*(\gamma,\eta) = (\delta\gamma, \ \delta\eta - \gamma).$$

Proof. By the induced riemannian metric on T^*_M, we see that $\|\omega\| = 1$ and $\omega \perp \hat{E}_\omega$. Thus, $\alpha + \delta\beta - < \alpha + \delta\beta, \omega >_{T^*_M}\omega$ means the $\hat{E}_\omega$-component of $\alpha + \delta\beta$ and hence $d\omega^{-1}$ can be applied. It is not hard to verify the above equalities, if we notice that

$< f\omega, \alpha >_o = << \omega, \alpha >_{T^*_M}, f >_o$, $\quad < \delta\beta, f\omega >_o = << \delta\beta, \omega >_{T^*_M}, f >_o$

$< \alpha - < \alpha,\omega >_{T^*_M}\omega, d\omega \lrcorner v >_o = < \alpha, d\omega \lrcorner v >_o$.

8.3.3 Lemma $\square = AA^* + B^*B$ <u>is elliptic and</u> $\mathrm{Ker}\,\square = \{0\}$.

Proof. We have easily that $\square(\alpha,\beta) = ((d\delta + \delta d)\alpha + \alpha, \ (d\delta + \delta d)\beta + \beta)$. Thus, we see that $\square$ is elliptic and $\mathrm{Ker}\,\square = \{0\}$.

Now, we can apply 4.5.5, 5.1.5 and 6.3.2 and get a strong ILH-Lie subgroup $\mathcal{D}'_\omega$ of $\Gamma_*(1_M)*\mathcal{D}$ with the Lie algebra Γ_ω, where $\mathcal{D}'^k_\omega$ is defined for $k \in N(\dim M + 7)$.

Let $\mathcal{D}_\omega = \{(f,\varphi) \in \Gamma_*(1_M)*\mathcal{D} \ : \ f\varphi^*\omega = \omega \}$, $\mathcal{D}^k_\omega = \{(f,\varphi) \in \Gamma^k_*(1_M)*\mathcal{D}^k \ : \ f\varphi^*\omega = \omega \}$. Then, $\mathcal{D}_\omega$, $\mathcal{D}^k_\omega$ are closed subgroups of $\Gamma_*(1_M)*\mathcal{D}$, $\Gamma^k_*(1_M)*\mathcal{D}^k$ respectively and by the same reasoning as in previous sections, it is easy to see that $\mathcal{D}'_\omega \subset \mathcal{D}_\omega$, $\mathcal{D}'^k_\omega \subset \mathcal{D}^k_\omega$.

Now, we define a mapping $\Psi : \Gamma_*(1_M)*\mathcal{D} \longmapsto \Gamma(T^*_M) \oplus \Gamma(\Lambda^2_M)$ by $\Psi(f,\varphi) = (f\varphi^*\omega, \ df\wedge \varphi^*\omega + f\varphi^*d\omega)$. We have then the following :

8.3.4 Lemma <u>The mapping Ψ can be extended to a smooth mapping of $\Gamma_*^{k+1}(1_M)*\mathcal{D}^{k+1}$ into $\Gamma^k(T_M^*) \oplus \Gamma^k(\Lambda_M^2)$ for every</u> $k \geq \dim M + 5$.

Proof. It is easy to see that the mapping $\wedge : \Gamma(\Lambda_M^p) \times \Gamma(\Lambda_M^q) \mapsto \Gamma(\Lambda_M^{p+q})$, $(\alpha,\beta) \mapsto \alpha \wedge \beta$, can be extended to a bounded bi-linear operator of $\Gamma^k(\Lambda_M^p) \times \Gamma^k(\Lambda_M^q)$ into $\Gamma^k(\Lambda_M^{p+q})$, where we use the conventions $\Lambda_M^0 = 1_M$, $\Lambda_M^1 = T_M^*$. Therefore, to prove this lemma is to prove the following :

8.3.5 Lemma <u>For an arbitrarily fixed</u> $\alpha \in \Gamma(\Lambda_M^p)$, <u>define a mapping</u> $\Phi : \mathcal{D} \mapsto \Gamma(\Lambda_M^p)$ <u>by</u> $\Phi(\varphi) = \varphi^*\alpha$. <u>Then, Φ can be extended to a smooth mapping of $\mathcal{D}^{k+1}$ into $\Gamma^k(\Lambda_M^p)$ for every</u> $k \geq \dim M + 5$.

Proof. It is easy to see that for any neighborhood $\widetilde{W}$ of e of $\mathcal{D}^{k+1}$, the union $\cup\{ \widetilde{W}g : g \in \mathcal{D} \} = \mathcal{D}^{k+1}$. Moreover, for every $g \in \mathcal{D}$, the mapping $g^* : \Gamma^k(\Lambda_M^p) \mapsto \Gamma^k(\Lambda_M^p)$ is a bounded linear operator, hence smooth. Therefore, we have only to prove that the mapping $\Phi : \widetilde{W} \mapsto \Gamma^k(\Lambda_M^p)$ is smooth.

Let Exp be the exponential mapping defined by a smooth connection on M. Let JT_M be the first jet bundle of T_M. For every element $Z \in JT_M$, we have a local vector field v such that $(j^1 v)(x) = Z$, where x is the base point of Z, and $(j^1 v)(x)$ means the 1-st jet of v at x. The mapping $y \mapsto \operatorname{Exp} v(y)$ defines a local diffeomorphism of M, if Z is very small, that is, there exists a relatively compact tubular neighborhood W^1 of the zero section of JT_M such that every $Z \in W^1$ defines a local diffeomorphism of M by the above manner. This local diffeomorphism will be denoted by φ_Z. This is a smooth diffeomorphism of a neighborhood of x (base point of Z) onto a neighborhood of $\operatorname{Exp} pZ$, where $p : JT_M \mapsto T_M$ be the natural projection.

Let $\pi : JT_M \mapsto M$ be the projection. We define a mapping $\nu : W^1 \mapsto \Lambda_M^p$ by $\nu(Z) = (\varphi_Z^* \alpha)(\pi Z)$. We see easily that this is well-defined and a smooth mapping.

Let $\Gamma(W^1) = \{ v \in \Gamma(T_M) : (j^1 v)(x) \in W^1$ for all $x \in M \}$, and let $\Gamma^k(W^1) = \{ v \in \Gamma^k(T_M) : (j^1 v)(x) \in W^1$ for all $x \in M \}$, $k \geq [\frac{1}{2}\dim M] + 2$. Then, we see that $\Phi(v)(x) = \nu(j^1 v(x))$. Thus, by 2.1.3 and 5.1.2, we see that Φ can be extended to a smooth mapping of $\Gamma^{k+1}(W^1)$ into $\Gamma^k(\Lambda_M^p)$.

Now, we will go back to the situation of 8.3.4 Lemma. The derivative of Ψ at e is equal to A, that is, $(d\Psi)_e(f,u) = A(f,u)$. Let E_1^k be the kernel of the mapping $B : \Gamma^k(T_M^*) \oplus \Gamma^k(\Lambda_M^2) \longmapsto \Gamma^{k-1}(\Lambda_M^2) \oplus \Gamma^{k-1}(\Lambda_M^3)$. Then, by the definition of Ψ and 8.3.4, Ψ is a smooth mapping of $\Gamma^{k+1}(1_M)*\mathcal{D}^{k+1}$ into E_1^k. Moreover, 8.3.3 Lemma shows that A is a homomorphism of $\Gamma^{k+1}(1_M) \oplus \Gamma^{k+1}(T_M)$ onto E_1^k. Thus, we can use the implicit function theorem of Hilbert manifolds, and hence we see that $\mathcal{D}_\omega^k$ is a smooth Hilbert submanifold of $\Gamma_*^k(1_M)*\mathcal{D}^k$. Thus, we get the following by the same reasoning as in the previous sections :

8.3.6 Theorem $\mathcal{D}_\omega$ <u>is a strong ILH-Lie subgroup of</u> $\mathbb{\Gamma}_*(1_M)*\mathcal{D}$ <u>and a closed subgroup.</u>

<u>Remark 1</u>. It is not difficult to see that Ψ is an C^∞ILHC^2-normal mapping. Thus, we can get the same result by using the implicit function theorem in §III.

<u>Remark 2</u>. There is a natural projection $\tilde{\pi} : \mathbb{\Gamma}_*(1_M)*\mathcal{D} \longmapsto \mathcal{D}$. This is obviously a homomorphism and can be extended to a smooth projection $\tilde{\pi}$ of $\Gamma_*^k(1_M)*\mathcal{D}^k$ onto $\mathcal{D}^k$. The restricted mapping $\tilde{\pi} : \mathcal{D}_\omega \longmapsto \mathcal{D}$ is clearly a monomorphism. However, the image $\tilde{\pi}\mathcal{D}_\omega$ (i.e. the ordinarly contact transformation group) is <u>not</u> a strong ILH-Lie subgroup of $\mathcal{D}$. The reason is the following :

Obviously, $(d\tilde{\pi})_e\mathbb{\Gamma}_\omega = \mathbb{\Gamma}_\omega(T_M)$. However, 8.3.1 Lemma shows that

$$\mathbb{\Gamma}_\omega = \{ -(dh \lrcorner \xi_\omega),\ h\xi_\omega + d\omega^{-1}(dh - (dh \lrcorner \xi_\omega)\omega) : h \in \mathbb{\Gamma}(1_M) \}$$

$$\mathbb{\Gamma}_\omega(T_M) = \{ h\xi_\omega + d\omega^{-1}(dh - (dh \lrcorner \xi_\omega)\omega) : h \in \mathbb{\Gamma}(1_M) \}.$$

Moreover, Γ_ω^k (the closure of $\mathbb{\Gamma}_\omega$ in $\Gamma^k(1_M) \oplus \Gamma^k(T_M)$) is given by

$$\{ -(dh \lrcorner \xi_\omega),\ h\xi_\omega + d\omega^{-1}(dh - (dh \lrcorner \xi_\omega)\omega) : h \in \Gamma^{k+1}(1_M) \}.$$

The point is that h is contained in $\Gamma^{k+1}(1_M)$ and not in $\Gamma^k(1_M)$. Thus, we see that $(d\tilde{\pi})_e\Gamma_\omega^k$ is not closed in $\Gamma^k(T_M)$.

VIII.4 Strictly contact transformations. (The proof of Theorem I.)

Let M be a 2m + 1 dimensional, smooth and closed manifold with a smooth contact form ω. Let ξ_ω be the vector field on M defined by $d\omega \lrcorner \xi_\omega \equiv 0$, $\omega \lrcorner \xi_\omega \equiv 1$. ξ_ω will be called a characteristic vector field of ω. We call ω a <u>regular contact form</u>, if ξ_ω induces a free action of the circle group S^1. Namely, there is a free action ρ of S^1 on M such that $\frac{d}{d\theta}\Big|_{\theta=0} \rho(\theta)x = \xi_\omega(x)$.

Let $\mathcal{D}_{s\omega} = \mathcal{D}_\omega \cap \{1\}*\mathcal{D}$. Another word, $\mathcal{D}_{s\omega} = \{ \varphi \in \mathcal{D} : \varphi^*\omega = \omega \}$. We call $\mathcal{D}_{s\omega}$ a strictly contact transformation group.

Since $\mathcal{L}_{\xi_\omega}\omega \equiv 0$, we see that $\rho(S^1)$ is contained in $\mathcal{D}_{s\omega}$.

8.4.1 Lemma $\mathcal{D}_{s\omega} = \{ g \in \mathcal{D}_\omega : g*(1,\rho(\theta)) = (1,\rho(\theta))*g$ for any $\theta \in S^1 \}$.

Proof. If $\varphi \in \mathcal{D}$ satisfies $\varphi^*\omega = \omega$, then $d\varphi\xi_\omega = \xi_\omega$. Thus, $\varphi\rho(\theta) = \rho(\theta)\varphi$. It follows $(1,\varphi)*(1,\rho(\theta)) = (1,\rho(\theta))*(1,\varphi)$. Conversely, let $(\mu,\varphi) \in \mathcal{D}_\omega$ with the property $(\mu,\varphi)*(1,\rho(\theta)) = (1,\rho(\theta))*(\mu,\varphi)$. Then, $\mu(\rho(\theta)(x)) = \mu(x)$ and $\varphi\rho(\theta) = \rho(\theta)\varphi$. This implies $d\varphi\xi_\omega = \xi_\omega$. Thus, $\frac{1}{\mu} = (\frac{1}{\mu}\omega)\lrcorner\xi_\omega = \varphi^*\omega\lrcorner\xi_\omega = \omega\lrcorner d\varphi\xi_\omega = 1$. Hence, $\mu = 1$ and $(\mu,\varphi) \in \mathcal{D}_{s\omega}$.

Now, by 7.1.4 Theorem, we have the following :

8.4.2 Theorem $\mathcal{D}_{s\omega}$ <u>is a strong ILH-Lie subgroup of</u> $\mathcal{D}_\omega$.

<u>Remark</u> $\mathcal{D}^k_{s\omega}$ is defined for $k \geq \dim M + 7$.

VIII.5 Finite codimensional subgroups of $\mathcal{D}_\Omega$ and $\mathcal{D}_{dV}$.

Let M be a 2m-dimensional smooth closed manifold with a smooth symplectic form Ω. Recall that $\Gamma_\Omega(T_M)$ is the totality of $u \in \Gamma(T_M)$ such that $\Omega \lrcorner u$ is a closed form. Let $\Gamma_\partial(T_M) = \{ u \in \Gamma_\Omega(T_M) : \Omega \lrcorner u$ is an exact form $\}$. Then, $\Gamma_\partial(T_M)$ is an ideal of $\Gamma_\Omega(T_M)$ because of the identity

$$\{\alpha,\beta\} = d(\alpha \lrcorner \beta^{\#}) + d\alpha \lrcorner \beta^{\#} - d(\beta \lrcorner \alpha^{\#}) - d\beta \lrcorner \alpha^{\#} + d(\Omega \lrcorner \alpha^{\#} \lrcorner \beta^{\#}),$$

where $\alpha^{\#}$ is defined by $\alpha = \Omega \lrcorner \alpha^{\#}$, and $\{\alpha,\beta\} = - \Omega \lrcorner [\alpha^{\#}, \beta^{\#}]$.

Since M is closed and $\Gamma_\Omega(T_M)/\Gamma_\partial(T_M) = H^1(M)$, we see that $\Gamma_\partial(T_M)$ is of finite codimension. Let $e_1,\dots,e_\nu$ be a linear basis of harmonic 1-forms with respect to a smooth riemannian metric on M. Then, $\Gamma_\partial(T_M)$ is perpendicular to every $e_i^{\#}$, i.e. $\int_M \langle u, e_i^{\#} \rangle dV = 0$ for any $u \in \Gamma_\partial(T_M)$. Therefore, by Theorem 1 in [31] together with Lemma 3 [31], we see that $\tilde{\Gamma}^k_\partial(T_M) = \{ dR_g \Gamma^k_\partial(T_M) : g \in \mathcal{D}^k \}$ is a smooth involutive distribution on $\mathcal{D}^k$ for any $k \geq \dim M + 7$. Hence we can apply 7.2.4. Moreover, since $\tilde{\Gamma}^k_\partial(T_M)$ is smooth, we have the following stronger result by the same method of 7.2.4 Theorem :

8.5.1 Theorem <u>There is a strong ILH-Lie subgroup</u> $\mathcal{D}_\partial$ <u>of</u> $\mathcal{D}_\Omega$ <u>with the Lie algebra</u> $\Gamma_\partial(T_M)$.

Let M be an arbitraly dimensional smooth closed manifold with a smooth volume element dV, i.e. an n-form, n = dimM. We consider the Lie algebra $\Gamma_{dV}(T_M) = \{ u \in \Gamma(T_M) : \operatorname{div} u = 0 \}$.

For any $u \in \Gamma_{dV}(T_M)$, we define an n-1-form $dV \lrcorner u$ by the following manner : Using a standard local coordinate such that $dV = dx^1 \wedge \cdots \wedge dx^n$, $dV \lrcorner u$ is defined by

$$dV \lrcorner u = \sum_{i=1}^{n} (-1)^{i-1} u^i \, dx^1 \wedge \cdots \wedge \check{dx}^i \wedge \cdots \wedge dx^n ,$$

where $u = \sum_{i=1}^{n} u^i \frac{\partial}{\partial x^i}$. This is well-defined and has an independent meaning from the choice of local coordinate as far as the local coordinate is such that dV =

$dx^1 \wedge \cdots \wedge dx^n$. $dV \lrcorner u$ is a closed form, as $\operatorname{div} u = 0$. Conversely, for any closed $n-1$ form, we can make a vector field contained in $\Gamma_{dV}(T_M)$.

Let $\Gamma_\delta(T_M) = \{ u \in \Gamma_{dV}(T_M) : dV \lrcorner u$ is an exact form $\}$. Then, $\Gamma_\delta(T_M)$ is an ideal of $\Gamma_{dV}(T_M)$, because of the identity $d(dV \lrcorner u \lrcorner v) = dV \lrcorner [u,v]$ for $u, v \in \Gamma_{dV}(T_M)$. Since $\Gamma_{dV}(T_M)/\Gamma_\delta(T_M) = H^{n-1}(M)$, $\Gamma_\delta(T_M)$ is finite codimensional, and by the same manner as above, we have the following :

8.5.2 Theorem <u>There is a strong ILH-Lie subgroup $\mathcal{D}_\delta$ of $\mathcal{D}_{dV}$ such that the Lie algebra of $\mathcal{D}_\delta$</u> is $\Gamma_\delta(T_M)$.

VIII.6 Sections of an involutive distribution.

Let F be a smooth involutive distribution on M and $\Gamma(F)$ the space of all smooth sections of F. $\Gamma(F)$ is, obviously, a subalgebra of $\Gamma(T_M)$. Let T_M/F be the factor bundle and π the natural projection of T_M onto T_M/F. This π induces naturally a linear mapping $\Gamma(T_M)$ onto $\Gamma(T_M/F)$, and $\Gamma(F)$ is the kernel of $\pi : \Gamma(T_M) \mapsto \Gamma(T_M/F)$. π can be regarded as a differential operator of order 0, and using smooth riemannian inner products on T_M, T_M/F, the formal adjoint operator $\pi^* : \Gamma(T_M/F) \mapsto \Gamma(T_M)$ is also a differential operator of order 0 such that $\pi\pi^*$ is elliptic (i.e. an isomorphism). Hence, we can apply 7.1.2 Corollary. There is a strong ILH-Lie subgroup $\mathcal{D}'_F$ of $\mathcal{D}$ with the Lie algebra $\Gamma(F)$.

Let F_x be the fibre of F at x and let $\mathcal{D}_F = \{ \varphi \in \mathcal{D} : d\varphi(F_x) = F_x$ for any $x \in M \}$. Then, it is not hard to see that $\mathcal{D}'_F$ is the connected component of $\mathcal{D}_F$ containing the identity.

Now, what we get is the following :

8.6.1 Theorem $\mathcal{D}_F$ <u>is a strong ILH-Lie subgroup of</u> $\mathcal{D}$.

<u>Remark</u> For the precise proof, we have to show at first that $\mathcal{D}^k_F$ is a smooth Hilbert manifold by using the exponential mapping with respect to a connection under which F is parallel.

Now, we assume that F gives a fibration of M with a compact fibre. Then, the fibre preserving diffeomorphism $\mathcal{D}_{\mathcal{F}}$ is a strong ILH-Lie group. (Cf. § II.) Let N be the base manifold of this fibre bundle. Then, 3.4.2 Theorem shows that a neighborhood of the identity of $\mathcal{D}_{\mathcal{F}}$ is coordinatized by the direct product of a neighborhood of e of $\mathcal{D}_{\mathbf{F}}$ and a neighborhood of e of $\mathcal{D}(N)$, because $\mathcal{D}_F$ = $\mathcal{D}_{[\mathcal{F}]}$ in the notion of 3.4.2 Theorem. Naturally, we have an exact sequence

$$0 \to \mathcal{D}_F \longrightarrow \mathcal{D}_{\mathcal{F}} \longrightarrow \mathcal{D}(N).$$

VIII.7 The group $\mathcal{D}_{\omega+\alpha}$ and the factor set $\mathcal{D}_{s\omega}\backslash\mathcal{D}_{\omega+\alpha}$.

Let M be a 2m+1-dimensional smooth closed manifold with a regular contact form ω. Then, using the characteristic vector field ξ_ω, we have a smooth fibering $\mathcal{F}$ of M with the fibre S^1. Let N be the factor space M/S^1. Since $\mathcal{L}_{\xi_\omega} d\omega = 0$, $d\omega$ can be regarded as a symplectic form on N. By 3.4.2 Theorem together with the above exact sequence, we see that $\tilde{\pi}^{-1}(\mathcal{D}_{d\omega}(N))$ is a strong ILH-Lie subgroup of $\mathcal{D}_{\mathcal{F}}$, by using the implicit function theorem.

Now, let $\mathcal{D}_{\omega+\alpha} = \{ \varphi \in \mathcal{D} : \varphi^*\omega = \omega + \alpha,\ \ \alpha$ is any closed 1-form $\}$. Since $\varphi^*\omega = \omega + \alpha$ is equivalent to $\varphi^* d\omega = d\omega$, we see that $d\varphi\xi_\omega = \tau\xi_\omega$, where τ is a smooth function. Thus, we have $\mathcal{D}_{\omega+\alpha} \subset \mathcal{D}_{\mathcal{F}}$ and hence $\mathcal{D}_{\omega+\alpha} = \tilde{\pi}^{-1}(\mathcal{D}_{d\omega}(N))$. Therefore, we have the following :

8.7.1 Theorem $\mathcal{D}_{\omega+\alpha}$ <u>is a closed strong ILH-Lie subgroup of</u> $\mathcal{D}_{\mathcal{F}}$.

Let $\mathcal{D}_{\omega+\alpha,o}$ be the totality of orientation preserving diffeomorphisms in $\mathcal{D}_{\omega+\alpha}$. Since $\varphi^*\omega \wedge \varphi^*(d\omega)^m = \omega \wedge (d\omega)^m$ (i.e. orientation preserving) for any $\varphi \in \mathcal{D}_{s\omega}$, we see that $\mathcal{D}_{s\omega} \subset \mathcal{D}_{\omega+\alpha,o}$.

Let $\mathbb{V} = \{ \omega + \alpha \in \Gamma(T^*_M) : d\alpha = 0,\ \int_{S^1} (\rho(\theta)^*\alpha) \lrcorner \xi_\omega \, d\theta = 0,\ \alpha \lrcorner \xi_\omega > -1 \}$, and

$\mathbb{F} = \{ \alpha \in \Gamma(T^*_M) : d\alpha = 0, \int_{S^1} (\rho(\theta)^*\alpha) \lrcorner \xi_\omega \, d\theta = 0 \}$. Then, $\mathbb{V}$ is an open contractible subset of $\omega + \mathbb{F}$.

The purpose of this section is to show the following :

8.7.2 Theorem $\mathcal{D}_{s\omega} \backslash \mathcal{D}_{\omega+\alpha,o}$ is homeomorphic to $\mathbb{V}$ and $\mathcal{D}_{\omega+\alpha,o}$ is a principal fibre bundle over $\mathbb{V}$ with the fibre $\mathcal{D}_{s\omega}$. Hence $\mathcal{D}_{\omega+\alpha,o}$ is homeomorphic to the direct product $\mathbb{V} \times \mathcal{D}_{s\omega}$.

This will be proved in the several lemmas below.

8.7.3 Lemma Every element of $\mathbb{V}$ is a regular contact form on M and $\varphi^*\mathbb{V} = \mathbb{V}$ for any $\varphi \in \mathcal{D}_{\omega+\alpha,o}$.

Proof. Let $\omega' = \omega + \alpha \in \mathbb{V}$. Then, $\omega' \wedge (d\omega')^m = \omega' \wedge (d\omega)^m$. Thus, $(\omega + \alpha) \wedge (d\omega)^m \neq 0$, if and only if $(\omega + \alpha) \lrcorner \xi_\omega \neq 0$, i.e. $1 + \alpha \lrcorner \xi_\omega \neq 0$, because $d\omega \lrcorner \xi_\omega = 0$. Thus, ω' is a contact form. The characteristic vector field $\xi_{\omega'}$ is given by

$$\xi_{\omega'} = \frac{1}{1 + \alpha \lrcorner \xi_\omega} \xi_\omega .$$

Therefore the integral curve of $\xi_{\omega'}$ is a circle. Let ρ' be the action of R (real numbers) on M generated by $\xi_{\omega'}$. So, every point $\rho'(t)x$ moves with the velocity vector $\xi_{\omega'}$. Thus, the condition $\int_{S^1} \rho(\theta)^*(\alpha \lrcorner \xi_\omega) d\theta = 0$ implies the time that the point $\rho'(t)x$ takes for a lap of integral curves is constant 1, and the converse is also true. Thus, ω' is a regular contact form.

For any $\varphi \in \mathcal{D}_{\omega+\alpha,o}$ and $\omega + \alpha \in \mathbb{V}$, we have $\varphi^*(\omega + \alpha) = \omega + \alpha_\varphi + \varphi^*\alpha$. Since this is a regular contact form, we have that $\omega + \alpha_\varphi + \varphi^*\alpha \in \mathbb{V}$ by the above argument

For an arbitrarily fixed element $\omega' \in \mathbb{V}$ (this may be different from ω), we denote by ρ' the free action of S^1 on M generated by $\xi_{\omega'}$.

8.7.4 Lemma $\mathbb{V}$ is equal to the subset $\{ \omega' + \beta \in \Gamma(T^*_M) : d\beta = 0, \int_{S^1} (\rho'(\theta)^*\beta \lrcorner \xi_{\omega'}) d\theta = 0, \beta \lrcorner \xi_{\omega'} > -1 \}$.

Proof. Put $\omega' = \omega + \alpha$. Obviously, $d\beta = 0$ if and only if $d(\alpha + \beta) = 0$, and

$\beta \lrcorner \xi_{\omega'} = \frac{\beta \lrcorner \xi_\omega}{1 + \alpha \lrcorner \xi_\omega} > -1$ if and only if $(\alpha + \beta) \lrcorner \xi_\omega > -1$. Moreover, we have

$$\int_{S^1} (\rho'(\theta)^*\beta) \lrcorner \xi_{\omega'} d\theta = \int_{S^1} \rho'(\theta)^*(\beta \lrcorner \xi_{\omega'}) d\theta = \int_{S^1} \rho(\theta')^*(\beta \lrcorner \xi_{\omega'}) d\theta,$$

where $\theta' = \theta'(\theta)$. Since $\frac{d\theta}{d\theta'} = 1 + \alpha \lrcorner \xi_\omega$, we see that

$$\int_{S^1} (\rho'(\theta)^*\beta) \lrcorner \xi_{\omega'} d\theta = \int_{S^1} (\rho(\theta)^*\beta) \lrcorner \xi_\omega \, d\theta.$$

For any fixed $\omega \in \mathbb{V}$, let $\Phi_\omega : \mathcal{D} \mapsto \Gamma(T^*_M)$ be the mapping defined by $\Phi_\omega(\varphi) = \varphi^*\omega$. By 8.3.5 Lemma, Φ_ω can be extended to a smooth mapping of $\mathcal{D}^{k+1}$ into $\Gamma^k(T^*_M)$. Therefore, the restricted mapping $\Phi_\omega : \mathcal{D}^{k+1}_{\omega+\alpha, o} \mapsto F^k$ is smooth, where F^k is the closure of $\mathbb{F}$ in $\Gamma^k(T^*_M)$, $k \geq \dim M + 5$.

8.7.5 Lemma <u>For any</u> $\omega \in \mathbb{V}$, <u>the mapping</u> $\Phi_\omega : \mathcal{D}_{\omega+\alpha, o} \mapsto \mathbb{F}$ <u>satisfies the conditions of the implicit function theorem (3.3.1 Theorem) on a neighborhood of the identity.</u>

Proof. Let ξ_ω be the characteristic vector field of ω, and $\rho(\theta)$ the free action of S^1 generated by ξ_ω. Consider the derivative $(d\Phi_\omega)_e : \Gamma_{\omega+\alpha}(T_M) \mapsto \mathbb{F}$, where $\Gamma_{\omega+\alpha}(T_M)$ is the tangent space of $\mathcal{D}_{\omega+\alpha}$ at e. Obviously, $\Gamma_{\omega+\alpha}(T_M) = \{ f\xi_\omega + v :$ $df + d\omega \lrcorner v =$ closed 1-form $\}$, where $v \in E_\omega = \{ X \in T_M : \omega \lrcorner X = 0 \}$.

Now, the derivative $(d\Phi_\omega)_e(f\xi_\omega + v)$ is equal to the Lie derivative $\mathcal{L}_{f\xi_\omega + v}\omega$ $= df + d\omega \lrcorner v$.

We define the same riemannian metric on M which is discussed in the page VIII. 4. By this riemannian metric, $\rho(\theta)$ is an isometry. Let α be a closed 1-form. Then, we can put $\alpha = dg + h$, where h is a harmonic 1-form and g is a function such that $\int_M g \, dV = 0$. If $\alpha \in \mathbb{F}$, then $\int_{S^1} \rho(\theta)^*(dg + h) \lrcorner \xi_\omega \, d\theta = 0$. Since $\int_{S^1} \rho(\theta)^* dg \lrcorner \xi_\omega \, d\theta = 0$ and $\rho(\theta)^* h = h$, we see that $h \lrcorner \xi_\omega = 0$. Thus, we can apply $d\omega^{-1}$ to h.

Define a mapping $J : \mathbb{F} \mapsto \Gamma_{\omega+\alpha}(T_M)$ by $J(dg + h) = g\xi_\omega + d\omega^{-1}(h)$. We see easily that $(d\Phi_\omega)_e J = \mathrm{id}$..

8.7.6 Lemma $\|J(dg + h)\|_{k+1} \leq C\|dg + h\|_k + D_k\|dg + h\|_{k-1}$ for any $k \geq 1$, where C, D_k are positive constant and C does not depend on k.

This lemma will be proved later. We assume for a while that this is true.

Put $p = J(d\Phi_\omega)_e : \Gamma_{\omega+\alpha}(T_M) \longmapsto \Gamma_{\omega+\alpha}(T_M)$. Then, $u = pu + (1-p)u$ gives the ILH-splitting of $\Gamma_{\omega+\alpha}(T_M)$. Since $(d\Phi_\omega)_e$ is a differential operator of order 1 with smooth coefficients, we see that $\|pu\|_k \leq C'\|u\|_k + D'_k\|u\|_{k-1}$ for any $k \geq \dim M + 6$ by using the above inequality for J. Thus, by the same manner as in the proof of 6.3.1 Lemma, we see that the above splitting is in fact an ILH-normal splitting. Therefore, to prove 8.7.5, we have only to check the conditions (a) and (b) in 3.3.1. However, the condition (a) is obvious in this case.

Recall 8.3.5 Lemma. We see that $\Phi_\omega(v)(x) = \nu(j^1 v(x))$, and ν is a smooth fibre preserving mapping of W^1 into T^*_M. By 2.1.3 and 5.1.2 Lemmas, we see that Φ_ω is a C^∞ILHC^2-normal mapping and hence so is $J\Phi_\omega$.

Proof of 8.7.6 Lemma.

We have $\|J(dg + h)\|_{k+1} = \|g\xi_\omega + d\omega^{-1}(h)\|_{k+1} \leq \|g\|_{k+1} + \|d\omega^{-1}(h)\|_{k+1}$. Since $\int_M g\,dV = 0$ and $d\omega : E_\omega \longmapsto \hat{E}_\omega$ (cf. 8.3.1 Lemma) is an isomorphism, we have

$$\|J(dg + h)\|_{k+1} \leq C(\|dg\|_k + \|h\|_{k+1}) + D_k(\|dg\|_{k-1}) + \|h\|_k).$$

On the other hand, let $\Box = d\delta + \delta d$. Then, $\Box + 1 : \Gamma(T^*_M) \longrightarrow \Gamma(T^*_M)$ is an elliptic differential operator. So, we have $\|h\|_{k-1} = \|(\Box + 1)h\|_{k-1} \geq C'\|h\|_{k+1} - D'_k\|h\|_k$. (Cf. 5.2.1 Lemma.) Use this inequality successively, then we get

$$\|h\|_{k+1} \leq C''\|h\|_k + D''_k\|h\|_{k-1} .$$

Insert this into the above inequality, and we get

$$\|J(dg + h)\|_{k+1} \leq C_1(\|dg\|_k + \|h\|_k) + D_{1,k}(\|dg\|_{k-1} + \|h\|_{k-1}).$$

Remark that $\square^{-1}d\delta(dg + h) = dg$, $(1 - \square^{-1}d\delta)(dg + h) = h$ and use the inequalities in the proof of 5.2.2 Theorem. Then we get

$$\|dg\|_k + \|h\|_k \leq C_2\|dg + h\|_k + D_{2,k}\|dg + h\|_{k-1} .$$

Inserting this into the above inequality, we have the desired result.

Now, we are ready to prove 8.7.2 Theorem.

We fix an element $\omega \in \mathbb{V}$. Remark that $\Phi_{\varphi^*\omega}(\psi) = \Phi_\omega(\varphi\psi)$. Then, by 8.7.5 Lemma together with the implicit function theorem, we have that $\Phi_\omega(\mathcal{D}_{\omega+\alpha,o})$ is an open subset of $\mathbb{V}$. Let $\omega' \in \mathbb{V}$ be any boundary point of $\Phi_\omega(\mathcal{D}_{\omega+\alpha,o})$. Then, $\Phi_{\omega'}(\mathcal{D}_{\omega+\alpha,o}) \cap \Phi_\omega(\mathcal{D}_{\omega+\alpha,o}) \neq \emptyset$, hence $\Phi_\omega(\mathcal{D}_{\omega+\alpha,o}) = \Phi_{\omega'}(\mathcal{D}_{\omega+\alpha,o})$. Thus, $\Phi_\omega(\mathcal{D}_{\omega+\alpha,o})$ is a closed subset of $\mathbb{V}$. Therefore, $\Phi_\omega(\mathcal{D}_{\omega+\alpha,o}) = \mathbb{V}$. By the implicit function theorem, we have that $\mathcal{D}_{s\omega} = \Phi_\omega^{-1}(\omega)$ is a strong ILH-Lie subgroup of $\mathcal{D}_{\omega+\alpha,o}$. Since $\Phi_\omega(\mathcal{D}_{s\omega}\varphi) = \Phi_\omega(\varphi)$, $\mathcal{D}_{\omega+\alpha,o}$ is a principal fibre bundle over $\mathbb{V}$ with the fibre $\mathcal{D}_{s\omega}$.

VIII.8 The factor set $\mathcal{D}_\Omega\backslash\mathcal{D}$.

Here, we consider the factor set $\mathcal{D}_\Omega\backslash\mathcal{D}$ by the similar method as above. Let $\mathbb{V}$ be the set of all symplectic 2-forms on M and $\mathbb{F}$ the space of all closed 2-forms on M. $\mathbb{V}$ is obviously an open subset of $\mathbb{F}$. For any $\Omega \in \mathbb{V}$, we consider a mapping $\Phi_\Omega : \mathcal{D} \mapsto \mathbb{F}$ defined by $\Phi_\Omega(\varphi) = \varphi^*\Omega$. Then, by 8.3.5 Lemma and its proof, we see that Φ_Ω is a C^∞ILHC^2-normal mapping in a neighborhood of the identity. (Cf. 2.1.3 and 5.1.2 Lemmas.)

On the other hand, the derivative $(d\Phi_\Omega)_e$ is given by the Lie derivative, that is, $(d\Phi_\Omega)_e u = d(\Omega \lrcorner u)$. Thus, putting $A = d(\Omega \lrcorner *)$, $B = d$, we have that $AA^* + B^*B$ is elliptic. (Cf. § VIII.2.). Therefore, it is not hard to verify that Φ_Ω satisfies all of the conditions of 3.3.1 Theorem. (See also 5.2.2 Theorem and 6.2.1 Lemma.)

Use the implicit function theorem, and we get the following :

8.8.1 Theorem <u>$\mathcal{D}_\Omega \backslash \mathcal{D}$ is homeomorphisc to an open, closed subset $\mathbb{V}'$ of $\mathbb{V}$ and $\mathcal{D}$ is a principal fibre bundle over $\mathbb{V}'$ with the fibre $\mathcal{D}_\Omega$.</u>

<u>Remark</u> The same procedure, we can apply to the factor set $\mathcal{D}_{dV} \backslash \mathcal{D}$. Then we get that $\mathcal{D}$ is homeomorphic to the direct product of $\mathcal{D}_{dV}$ and $\mathbb{V}$, where $\mathbb{V}$ is the totality of all volume elements of total volume 1. $\mathbb{V}$ is obviously a contractible set. (Cf. [28].)

VIII.9 The factor set $\mathcal{D}_{s\omega} \backslash \mathcal{D}_\omega$

Let ω be a regular contact form on M. For any $\tau \in \Gamma_*(1_M)$, we see that $\tau\omega$ is a contact form and $\mathcal{D}_{\tau\omega} = \mathcal{D}_\omega$. The characteristic vector field $\xi_{\tau\omega}$ is given by $\frac{1}{\tau}\xi_\omega - \{\frac{1}{\tau}\}_\omega$, where $\{\frac{1}{\tau}\}_\omega = d\omega^{-1}(d(\frac{1}{\tau}) - (d(\frac{1}{\tau}) \lrcorner \xi_\omega)\omega)$.

Let $\mathcal{J}$ be the totality of functions $\tau \in \Gamma_*(1_M)$ such that $\tau\omega$ is a regular contact form.

8.9.1 Lemma <u>$\mathcal{J}$ is a closed subset of</u> $\Gamma_*(1_M)$.

Proof. Let $\{\tau_n\}$ be a sequence in $\mathcal{J}$ converging to an element $\tau \in \Gamma_*(1_M)$. Then, the characeristic vector fields $\xi_{\tau_n\omega}$ converges to $\xi_{\tau\omega}$. We see that every $\xi_{\tau_n\omega}$ induces a free action ρ_n of S^1 onto M. Therefore, $\xi_{\tau\omega}$ induces an action ρ of S^1 onto M. However, the result in [35] shows that ρ and ρ_n are conjugate for sufficiently large n. This implies that ρ is also a free action. Hence $\tau\omega$ is a regular contact form.

Now, the purpose of this section is to prove the following :

8.9.2 Theorem $\mathcal{D}_{s\omega} \backslash \mathcal{D}_\omega$ <u>is homeomorphic to an open and closed subset of</u> $\mathcal{J}$.

This will be proved in several lemmas below.

We put $\tilde{\mathcal{J}} = \mathcal{J}\omega$.

For any $\omega' \in \tilde{\mathfrak{J}}$, we consider the mapping $\Phi_{\omega'} : \mathcal{D}_\omega \mapsto \Gamma(T^*_M)$ by $\Phi_{\omega'}(g,\varphi) = \frac{1}{g}\omega' = \varphi^*\omega'$. $\Phi_{\omega'}$ can be extended to a smooth mapping of $\mathcal{D}^k_\omega$ into $\Gamma^k(T^*_M)$ for any $k \geq \dim M + 7$, and obviously $\varphi^*\tilde{\mathfrak{J}} = \tilde{\mathfrak{J}}$ for any $(g,\varphi) \in \mathcal{D}_\omega$.

At first, we will give the proof of 8.9.2 by assuming the following :

8.9.3 Lemma <u>For any</u> $\tau \in \mathfrak{J}$ <u>and for any sequence</u> $\{\tau_n\} \subset \mathfrak{J}$ <u>converging to</u> τ, <u>there exist</u> $(g_n, \varphi_n) \in \mathcal{D}_\omega$ <u>such that</u> $(g_n, \varphi_n) \mapsto e$ <u>with</u> $n \to \infty$ <u>and</u> $\varphi^*_n\tau\omega = \tau_n\omega$ <u>for sufficiently large</u> n.

8.9.2 is an easy application of the above lemma. Remark that $\Phi_{\varphi^*\omega}(g,\psi) = (\varphi\psi)^*\omega$. We see easily that $\Phi_\omega(\mathcal{D}_\omega)$ is an open subset of $\tilde{\mathfrak{J}}$. Let $\tau\omega$ be a boundary point of $\Phi_\omega(\mathcal{D}_\omega)$. Then the above lemma shows that $\Phi_{\tau\omega}(\mathcal{D}_\omega) \cap \Phi_\omega(\mathcal{D}_\omega) \neq \emptyset$, hence $\Phi_{\tau\omega}(\mathcal{D}_\omega) = \Phi_\omega(\mathcal{D}_\omega)$. Thus, $\Phi_\omega(\mathcal{D}_\omega)$ is an open and closed subset of $\tilde{\mathfrak{J}}$.

Let $\tilde{\mathfrak{J}}' = \Phi_\omega(\mathcal{D}_\omega)$. $\Phi_\omega : \mathcal{D}_\omega \mapsto \tilde{\mathfrak{J}}'$ induces a continuous one to one mapping of $\mathcal{D}_{s\omega} \backslash \mathcal{D}_\omega$ onto $\tilde{\mathfrak{J}}'$. Again, the above lemma shows that this is a homeomorphism.

The Corollary in § 0 is easy to prove by 8.9.2 and 8.10.1 Theorem below.

Now, to complete the proof of 8.9.2, we have only to show the above lemma. This lemma would be trivial if one could apply the implicit function theorem. However, in this case we can not use it. The reason is the following : First, we do not know whether $\mathfrak{J}$ is a Frechet manifold or not. Secondly, if $\mathfrak{J}$ is a Frechet manifold, then the tangent space of $\mathfrak{J}$ at 1 should be $\mathbb{E} = \{dh \lrcorner \xi_\omega : h \in \Gamma(1_M)\}$, because

$$(d\Phi_\omega)_e(\, dh\lrcorner\xi_\omega,\ h\xi_\omega - d\omega^{-1}(dh - (dh\lrcorner\xi_\omega)\omega)) = -\, dh\lrcorner\xi_\omega \,.$$

Let E^k be the closure of $\mathbb{E}$ in $\Gamma^k(1_M)$. Then, $E^k = \{\, f \in \Gamma^k(1_M) : \int_{S^1}\rho(\theta)^*f\, d\theta = 0 \,\}$. The mapping $(d\Phi_\omega)_e$ can be extended to a continuous mapping of Γ^{k+1}_ω into E^k, but this is not surjective.

Thus, we must use other method.

Proof of 8.9.3 Lemma.

Let $\{\tau_n\}$ be a sequence in $\mathcal{T}$ converging to τ. Then, $\xi_{\tau_n \omega}$ converges to $\xi_{\tau\omega}$. Thus, Palais [35] shows that there is $\varphi_n \in \mathcal{D}$ such that $\varphi_n \mapsto e$ and $d\varphi_n \xi_{\tau\omega} = \xi_{\tau_n \omega}$ for sufficiently large n. We put $\omega'_n = \varphi^*_n \tau_n \omega$. ω'_n is also a regular contact form on M and $\xi_{\tau\omega} = \xi_{\omega'_n}$. Thus, both contact forms $\tau\omega$ and ω'_n induce the same free action of S^1 on M. Let N be the factor space M/S^1. Since $\mathcal{L}_{\xi_{\tau\omega}} d(\tau\omega) = 0$ and $\mathcal{L}_{\xi_{\omega'_n}} d\omega'_n = 0$, $d(\tau\omega)$ and $d\omega'_n$ can be regarded as symplectic forms on N. Since $\varphi_n \mapsto e$, we see that $d\omega'_n \mapsto d(\tau\omega)$.

Now, we use 8.8.1 Theorem. For a sufficiently large n, there is $\psi_n \in \mathcal{D}(N)$ such that $\psi_n \mapsto e$ and $\psi^*_n d(\tau\omega) = d\omega'_n$.

Let $\widetilde{\psi}_n$ be the lift of ψ_n. $\widetilde{\psi}_n : M \mapsto M$ can be defined for a sufficiently large n, and $\widetilde{\psi}_n \mapsto e$ with $n \mapsto \infty$. We see easily that $\widetilde{\psi}^{-1*}_n \omega'_n = \tau\omega + \alpha_n$, $d\alpha_n = 0$. Moreover, $\alpha_n \mapsto 0$ if $n \mapsto \infty$. Thus, $\widetilde{\psi}^{-1*}_n \varphi^*_n \tau_n \omega = \tau\omega + \alpha_n$. Since $\tau\omega + \alpha_n$ is a regular contact form, we have that

$$\int_{S^1} \rho(\theta)^* \alpha_n \lrcorner\, \xi_{\tau\omega} \, d\theta = 0,$$

where $\rho(\theta)$ is the action of S^1 generated by $\xi_{\tau\omega}$.

Now, we use 8.7.2 Theorem. There is $\zeta_n \in \mathcal{D}_{\tau\omega+\alpha}$ such that $\zeta_n \mapsto e$ and $\zeta^*_n \tau\omega = \tau\omega + \alpha_n$. Thus, $(\varphi_n \widetilde{\psi}^{-1}_n \zeta^{-1}_n)^* \tau_n \omega = \tau\omega$, hence $(\frac{\tau}{\tau_n}, \zeta_n \widetilde{\psi}_n \varphi^{-1}_n)$ is the desired sequence.

VIII.10 The factor set $\mathcal{D}_\omega \backslash \Gamma_*(1_M) * \mathcal{D}$.

Here, we consider the factor set $\mathcal{D}_\omega \backslash \Gamma_*(1_M) * \mathcal{D}$ by the same method as in § VIII.8.

Let $\mathbb{V}$ be the totality of pairs $(\omega, d\omega) \in \Gamma(T^*_M) \oplus \Gamma(\Lambda^2_M)$ such that ω is a contact form on M. Let $\mathbb{F}$ be the linear subspace of $\Gamma(T^*_M) \oplus \Gamma(\Lambda^2_M)$ consisting of all elements $(\alpha, d\alpha)$. $\mathbb{V}$ is obviously an open subset of $\mathbb{F}$. For any $(\omega, d\omega) \in \mathbb{V}$, we consider a mapping $\Phi_\omega : \Gamma_*(1_M)*\mathcal{D} \longmapsto \mathbb{F}$ defined by $\Phi_\omega(\tau,\varphi) = (\tau\varphi^*\omega,\ d\tau\wedge\varphi^*\omega + \tau\varphi^*d\omega\ )$. Then, by 8.3.4 and its proof, we see that Φ_ω is a C^∞ILHC^2-normal mapping. Moreover,

$$(d\Phi_\omega)_e(f,u) = (f\omega + d(\omega\lrcorner u) + d\omega\lrcorner u,\ d(f\omega) + d(d\omega\lrcorner u)).$$

Thus, putting $A = (d\Phi_\omega)_e$, $B(\alpha,\beta) = (d\alpha - \beta,\ d\beta)$, we have that $AA^* + B^*B$ is elliptic. Thus, it is easy to see that Φ_ω satisfies the conditions of 3.3.1 Theorem.

Use the implicit function theorem, and we see

8.10.1 Theorem $\mathcal{D}_\omega \backslash \Gamma_*(1_M)*\mathcal{D}$ <u>is homeomorphic to an open and closed subset</u> $\mathbb{V}'$ <u>of</u> $\mathbb{V}$ <u>and</u> $\Gamma_*(1_M)*\mathcal{D}$ <u>is a principal fibre bundle over</u> $\mathbb{V}'$ <u>with the fibre</u> $\mathcal{D}_\omega$.

Especially, for any ω' which is sufficiently near to ω, we can find $(\tau,\varphi) \in \Gamma_*(1_M)*\mathcal{D}$ such that $(\omega', d\omega') = (\tau\varphi^*\omega,\ d\tau\wedge\varphi^*\omega + \tau d\varphi^*\omega)$, that is, $\omega' = \tau\varphi^*\omega$. This implies that any deformation of contact form is locally trivial up to function factor. Therefore, the above result contains the result in Gray [13] Theorem 5.2.1.

§ IX Primitive transformation groups

The purpose of this chapter is to prove Theorem H in § 0. Recall the definition of $\mathfrak{g}$ there. We will discuss the case $\dim \mathfrak{g} < \infty$ and the case $\dim \mathfrak{g} = \infty$ separately.

IX.1 Finite dimensional Lie algebras.

Let G be a closed subgroup of $\mathcal{D}$ such that $\mathfrak{g}$ is a finite dimensional subalgebra of $\Gamma(T_M)$. (C.f. 1.4.1.) Then, the closure of $\mathfrak{g}$ in $\Gamma^k(T_M)$ is equal to $\mathfrak{g}$ itself and therefore $\tilde{\mathfrak{g}} = \{dR_g\mathfrak{g} : g \in \mathcal{D}^k\}$ is a smooth involutive distribution of $\mathcal{D}^k$ for any $k \geq \dim M + 5$. (C.f. (G,7) and Proposition A in [31].) Thus, using Frobenius theorem in Hilbert manifolds, we take the maximal integral submanifold G' through e. This is a subgroup of $\mathcal{D}^k$ for any k, hence $G' \subset \mathcal{D}$. Therefore, the group operations are smooth and hence G' is a Lie group with the Lie algebra $\mathfrak{g}$. By 7.1.5 Theorem, we see that G' is the connected component of G under the LPSAC-topology.

This is the first half of Theorem H. However, we can get a more precise result by considering Frobenius theorem. We will discuss it in slightly general manner.

Let G be a strong ILB-Lie group with the Lie algebra $\mathfrak{g}$. Suppose $\mathfrak{h}$ is a finite dimensional subalgebra of $\mathfrak{g}$. Then, $\mathfrak{h}$ is a closed subspace of $\mathfrak{g}^k$ for every $k \in N(d)$. In particular, $\mathfrak{h}$ is a closed subspace of $\mathfrak{g}^d$ and hence there is a closed subspace $\mathfrak{m}^d$ of $\mathfrak{g}^d$ such that $\mathfrak{g}^d = \mathfrak{h} \oplus \mathfrak{m}^d$. For every $u \in \mathfrak{g}^d$, we have $u = v + w$ in accordance with that splitting. Since $v \in \mathfrak{h} \subset \mathfrak{g}$, if $u \in \mathfrak{g}^k$, then $w \in \mathfrak{m}^d \cap \mathfrak{g}^k = \mathfrak{m}^k$. Thus, we have a complementary chain $\{\mathfrak{m}, \mathfrak{m}^k, k \in N(d)\}$, where $\mathfrak{m} = \mathfrak{m}^d \cap \mathfrak{g}$.

9.1.1 Theorem <u>Notations being as above, there is a strong ILB-Lie subgroup</u> H <u>of</u> G <u>with the Lie algebra</u> $\mathfrak{h}$.

Proof. Let $\xi : U \cap \mathfrak{g} \to \tilde{U} \cap G$ be an ILB-coordinate mapping. Since $\tilde{\mathfrak{h}} =$

$\{ dR_g\mathfrak{h} : g \in G^k\}$ is a smooth distribution on G^k, this defines a smooth involutive distribution on $U \cap \mathfrak{g}^k$ through the mapping ξ.

Let $\tilde{\mathfrak{h}}_u$ be the fibre of that distribution at $u \in U \cap \mathfrak{g}^k$, that is, $\tilde{\mathfrak{h}}_u = d\xi^{-1}dR_{\xi(u)}\mathfrak{h}$. Let π be the projection of $\mathfrak{g}$ onto $\mathfrak{h}$ in accordance with the splitting $\mathfrak{g} = \mathfrak{h} \oplus \mathfrak{m}$. Then, π can be extended to the projection of $\mathfrak{g}^k$ onto $\mathfrak{h}$. Since $\dim \mathfrak{h} < \infty$, there are open convex neighborhoods V, W of zeros of $\mathfrak{h}$, $\mathfrak{m}^d$ such that $V \times W \subset U$ and that the restriction $\pi | \tilde{\mathfrak{h}}_u$ is an isomorphism of $\tilde{\mathfrak{h}}_u$ onto $\mathfrak{h}$ for every $u \in V \times W \cap \mathfrak{m}^k$, $k \in N(d)$. Thus, we have a mapping $\Phi : \mathfrak{h} \times V \times W \cap \mathfrak{m} \longrightarrow \mathfrak{m}$ which describes the distribution on $V \times W \cap \mathfrak{m}^k$. Namely, Φ has the following properties :

(i) Φ is linear with respect to the first variable $h \in \mathfrak{h}$.

(ii) $\Phi(h,0,0) = 0$.

(iii) Φ can be extended to a smooth mapping of $\mathfrak{h} \times V \times W \cap \mathfrak{m}^k$ into $\mathfrak{m}^k$ for every $k \in N(d)$.

(iv) $\{ (h,\Phi(h,x,y)) : h \in \mathfrak{h} \} = \tilde{\mathfrak{h}}_{(x,y)}$ for every $(x,y) \in V \times W \cap \mathfrak{m}^k$.

Recall the arguments in § VI. For any $x \in V$, we denote by $\bar{y}(t,x,y)$ the solution of $\frac{d}{dt}y(t) = \Phi(x, tx, y(t))$ with the initial condition $y(0) = y$. By takeing V, W very small, we may assume that $\bar{y}(t,x,y)$ is defined for every $t \in [0,1]$ and for every $(x,y) \in V \times W$ as a solution in $\mathfrak{m}^d$.

Let $\bar{y}(t,x,y)$ be a solution in $\mathfrak{m}^d$. Then, by the above property (iii), we see that if $\bar{y}(t,x,y) \in \mathfrak{m}^k$ (resp. $\mathfrak{m}$) for some $t \in [0,1]$, then $\bar{y}(t,x,y) \in \mathfrak{m}^k$ (resp. $\mathfrak{m}$) for every $t \in [0,1]$.

Put $\Psi(x,y) = \bar{y}(1,x,y)$ for $(x,y) \in V \times W \cap \mathfrak{m}$. Then, by the differentiability of the solutions with respect to the initial conditions shows that Ψ can be extended to a smooth mapping of $V \times W \cap \mathfrak{m}^k$ into $U \cap \mathfrak{g}^k$ for every $k \in N(d)$. Since the equation is a differential equation of first order, we see that Ψ is a smooth diffeomorphism of $V \times W \cap \mathfrak{g}^k$ onto an open neighborhood U'^k of 0 of $U \cap \mathfrak{g}^k$.

By the property of $\bar{y}(t,x,y)$ stated above, we see that $U'^k = U'^d \cap \mathfrak{g}^k$, that is, Ψ is a smooth diffeomorphism of $V \times W \cap \mathfrak{g}^k$ onto an open neighborhood $\Psi(V \times W) \cap \mathfrak{g}^k$.

Thus, $\xi\Psi : V \times W \cap \mathfrak{M} \longmapsto G$ is an ILB-coordinate mapping such that $\xi\Psi(V \times \{u\})$ is an integral submanifold. Therefore, we have that there is a strong ILB-Lie subgroup H with the Lie algebra $\mathfrak{h}$.

Remark 1 If H is a closed subgroup of G, then taking V, W very small if necessary, we may assume that $H \cap \xi(\{0\} \times W) = \{e\}$. Then, the mapping $\xi : \{0\} \times W \cap \mathfrak{M} \longrightarrow G$ gives a local cross-section of $H\backslash G$ and gives a manifold structure on $H\backslash G$.

Remark 2 The next theorem refers to the closedness of the subgroup H.

A finite dimensional Lie group H is called a (CA)-group, if the image of the adjoint representation $Ad(H)$ of H on the Lie algebra $\mathfrak{h}$ is closed in $GL(\mathfrak{h})$. Then, the following is a special case of Theorem 1.1 of [27]:

9.1.2 Theorem If an ILB-Lie subgroup H of G is a finite dimensional (CA)-group with compact center. Then, H is always a closed subgroup of G.

We have following examples of (CA)-groups H with compact centers :

(a) Finite dimensional connected semi-simple Lie group with finite center. (Cf. Corollary 1.1 of [27].)

(b) Finite dimensional connected nilpotent Lie group with compact center. (Cf. Corollary 1.3 of [27].)

Some other examples are given in Theorem 1.2 of [27].

IX.2 Primitive transformation groups.

The purpose of this section is to show the following :

9.2.1 Theorem <u>If</u> G <u>satisfies the conditions (a) ~ (c) in Theorem H in § 0 and</u> $\dim \mathfrak{g} = \infty$, <u>then</u> G <u>is an open subgroup of one of</u> $\mathfrak{D}$, $\mathfrak{D}_{dV}$, $\mathfrak{D}_{\Omega}$, $\tilde{\mathfrak{D}}_{\omega}$ <u>and</u> $\mathfrak{g}$ <u>is equal with one of</u> $\Gamma(T_M)$, $\Gamma_{dV}(T_M)$, $\Gamma_{\Omega}(T_M)$, $\Gamma_{\omega}(T_M)$, <u>where</u> $\tilde{\mathfrak{D}}_{\omega} = \{ g \in \mathfrak{D} : g^*\omega = \tau\omega$ for a C^{∞}-function $\tau \}$ <u>and</u> M <u>is assumed to be orientable.</u>

The above theorem will be proved in the several lemmas below.

Now, start with the Lie algebra $\mathfrak{g}$ such that $\dim \mathfrak{g} = \infty$. Let $\mathfrak{g}_k(p) = \{ u \in \mathfrak{g} : (j^k u)(p) = 0 \}$, where $(j^k u)(p)$ means the k-th jet of u at p. We have then

$$\mathfrak{g} = \mathfrak{g}_{-1}(p) \supset \mathfrak{g}_0(p) \supset \mathfrak{g}_1(p) \supset \dots \supset \mathfrak{g}_k(p) \supset \dots ,$$

$$[\mathfrak{g}_k(p), \mathfrak{g}_{\ell}(p)] \subset \mathfrak{g}_{k+\ell}(p) \quad \text{and} \quad \dim \mathfrak{g}_k(p)/\mathfrak{g}_{k+1}(p) < \infty. \quad \text{(C.f. 1.4.1.)}$$

Since G is transitive on M (the condition (a)), this filtration does not depend on $p \in M$, that is, letting g be an element such that $g(p) = q$, we see $Ad(g)\mathfrak{g}_k(q) = \mathfrak{g}_k(p)$ for any $k \geq -1$, where $(Ad(g)u)(z) = dg\,u\,(g^{-1}(z))$.

For any $p \in M$, we consider $\mathfrak{g}/\mathfrak{g}_0(p)$. Since G is transitive, this defines a smooth distribution on M, and as $\mathfrak{g}$ is a Lie algebra, this distribution is involutive. Thus, by the primitivity, we see that $\dim \mathfrak{g}/\mathfrak{g}_0(p) = \dim M$.

9.2.2 Lemma <u>Let</u> $\mathfrak{J}(p) = \bigcap_k \mathfrak{g}_k(p)$. <u>If</u> $\dim \mathfrak{g}/\mathfrak{J}(p) < \infty$, <u>then</u> $\dim \mathfrak{g} < \infty$.

Proof. By the assumption, there is $m \geq 0$ such that $\mathfrak{g}_{m-1}(p) \supsetneq \mathfrak{g}_m(p) = \mathfrak{g}_{m+1}(p) = \dots = \mathfrak{J}(p)$. For any $p \in M$, there are $e_1, \dots, e_{\nu}$ such that $\bar{e}_1, \dots, \bar{e}_{\nu} \in \mathfrak{g}/\mathfrak{J}(p) = \mathfrak{g}/\mathfrak{g}_m(p)$ form a linear basis of $\mathfrak{g}/\mathfrak{g}_m(p)$, where $\bar{e}_i$ means $e_i + \mathfrak{g}_m(p)$.

On the other hand, to consider an element $u + \mathfrak{g}_m(p)$ is to consider $(j^m u)(p)$. Therefore, $\mathfrak{g}/\mathfrak{g}_m(p)$ can be regarded as a subspace of $J^m_p T_M$ (m-th jet space of vector fields at p), and $(j^m e_1)(p), \dots, (j^m e_{\nu})(p)$ form a basis of $\mathfrak{g}/\mathfrak{g}_m(p)$. Obviously, there exists an open neighborhood U_p of p such that $\{(j^m e_1)(q), \dots, (j^m e_{\nu})(q)\}$ is a basis of $\mathfrak{g}/\mathfrak{g}_m(q)$ for any $q \in U_p$. Thus, $\{ \mathfrak{g}/\mathfrak{g}_m(p) : p \in M \}$

gives a smooth subbundle of $J^m T_M$ and for any $u \in \mathfrak{J}$, there are smooth functions $\lambda_1, \ldots, \lambda_\nu$ on U_p such that $j^m u = \sum \lambda_i j^m e_i$.

On the other hand, for any fixed $z \in U_p$, we have $u - \sum \lambda_i(z) e_i \in \mathfrak{J}_m(z) = \mathcal{J}(z)$. This implies $(j^{m+1}u)(z) = \sum \lambda_i(z)(j^{m+1}e_i)(z)$, hence

$$(\partial_j \, j^m u)(z) = \sum \lambda_i(z)(\partial_j j^m e_i)(z)$$

for any x_j, where $(x_1, \ldots, x_n)$ is a local coordinate on U_p and we use the notation ∂_j instead of $\frac{\partial}{\partial x_j}$.

Take the derivative of $j^m u = \sum \lambda_i j^m e_i$ at z and compair with the above equality. Then, we have $(\partial_j \lambda_i)(z) = 0$, so that if $u \in \mathcal{J}(p)$, then $u = 0$ on the connected component of M containing p. Recall that M is a closed manifold, hence there are only finite number of components. Thus, $\dim \mathcal{J}(p) < \infty$, hence $\dim \mathfrak{J} < \infty$.

Thus, to assume $\dim \mathfrak{J} = \infty$ is to assume $\dim \mathfrak{J}/\mathcal{J}(p) = \infty$ for any $p \in M$. Let $\bar{\mathfrak{J}}_i(p) = \mathfrak{J}_i(p)/\mathcal{J}(p)$. Then, $\bar{\mathfrak{J}}_{-1} \supset \bar{\mathfrak{J}}_0(p) \supset \ldots \supset \bar{\mathfrak{J}}_k(p) \supset \ldots$ satisfies the conditions of filtered Lie algebras [17, III]. Obviously, $(\bar{\mathfrak{J}}_{-1}, \bar{\mathfrak{J}}_0(p))$ is a primitive filtered Lie algebra. Now, we define a topology for $\bar{\mathfrak{J}}_{-1}(p)$ by taking $\{\bar{\mathfrak{J}}_k(p)\}$ as a basis of neighborhoods of 0. We denote by $\mathfrak{a}(\mathfrak{J}, p)$ the completion of $\bar{\mathfrak{J}}_{-1}(p)$ by the above topology and $\mathfrak{a}_k(\mathfrak{J}, p)$ the closure of $\bar{\mathfrak{J}}_k(p)$ in $\mathfrak{a}(\mathfrak{J}, p)$. We see easily that $\mathfrak{a}(\mathfrak{J}, p)/\mathfrak{a}_k(\mathfrak{J}, p) = \bar{\mathfrak{J}}_{-1}(p)/\bar{\mathfrak{J}}_k(p) = \mathfrak{J}/\mathfrak{J}_k(p)$ for any k and that $\mathfrak{a}(\mathfrak{J}, p)$ is a primitive complete filtered Lie algebra, that is, a primitive TLA in the notation of [23].

Let F be the totality of formal power series $\sum a^i_\alpha x^\alpha \partial_i$ of vector fields. Let $F_k = \{ u \in F : u = \sum_{|\alpha| \geq k+1} a^i_\alpha x^\alpha \partial_i \}$. Define a topology on F by taking $\{F_k\}$ as a basis of neighborhoods of 0. Then, F is a complete filtered Lie algebra. Obviously, to consider $\mathfrak{a}(\mathfrak{J}, p)$ is to consider a closed subalgebra of F.

Let B be an open ball in R^n with the center 0 and with the natural coordinate. We can do the same procedure for any subalgebra $\mathfrak{g}$ of $\Gamma(T_B)$, that is, we can make $\mathfrak{a}(\mathfrak{g},0)$.

Now, according to the classification of primitive TLA [23][15] and [40], we see that the infinite dimensional primitive TLA which is made from a subalgebra of vector fields on a closed manifold is one of the following :

$$\mathfrak{a}(\Gamma(T_B),0),\quad \mathfrak{a}(\Gamma_{dV_o}(T_B),0),\quad \mathfrak{a}(\Gamma_{\Omega_o}(T_B),0),\quad \mathfrak{a}(\Gamma_{\omega_o}(T_B),0),$$

where $dV_o = dx_1\wedge\cdots\wedge dx_n$, $\Omega_o = \sum_{i=1}^{m} dx_i\wedge dx_{m+i}$, $\omega_o = dx_o + \sum_{i=1}^{m} x_i dx_{m+i}$.

(See also [17, III] and [41].) All other cases which appear in the classification are absorbed into the above cases or are elliptic, that is, it becomes a finite dimensional Lie algebra on a closed manifold. Though we do not give the precise reasoning of the above facts, it will be clear by the same method as in the proof of the next lemma.

9.2.3 Lemma <u>If</u> G <u>satisfies (a) ~ (c) and</u> $\dim\mathfrak{g}=\infty$, <u>then there is a smooth volume element</u> dV, <u>a smooth symplectic form</u> Ω <u>or a smooth contact form</u> ω <u>on</u> M <u>such that</u> G <u>is contained in one of</u> $\mathcal{D}$, $\mathcal{D}_{dV}$, $\mathcal{D}_\Omega$, $\tilde{\mathcal{D}}_\omega$. <u>Therefore,</u> $\mathfrak{g}$ <u>is contained in one of</u> $\Gamma(T_M)$, $\Gamma_{dV}(T_M)$, $\Gamma_\Omega(T_M)$, $\Gamma_\omega(T_M)$.

Proof. Remark first of all that $\mathfrak{a}(\mathfrak{g},q)\cong\mathfrak{a}(\mathfrak{g},p)$ for any $q\in M$, because G is transitive on M.

(1) The case $\mathfrak{a}(\mathfrak{g},p)\cong\mathfrak{a}(\Gamma(T_B),0)$: There is nothing to prove.

(2) The case $\mathfrak{a}(\mathfrak{g},p)\cong\mathfrak{a}(\Gamma_{dV_o}(T_B),0)$: Let $(x_1,\dots,x_n)$ be a coordinate system at p with the origin corresponding to p. Take Taylor expansion of $u\in\mathfrak{g}$ in the coordinate expressions and take the closure in F. Then, we get a Lie subalgebra $\mathfrak{a}(\mathfrak{g},(x_1,\dots,x_n),p)$ of F. Remark that $\mathfrak{a}(\mathfrak{g},p)\cong\mathfrak{a}(\Gamma_{dV_o}(T_B),0)$ implies that $\mathfrak{a}(\mathfrak{g},(x_1,\dots,x_n),p)=\mathfrak{a}(\Gamma_{dV_o}(T_B),0)$ for suitable choice of local coordinate.

Define an n-form dV_p at p by $dx_1\wedge\cdots\wedge dx_n$ using this coordinate. Recall that if $g\in G$ such that $g(q)=p$, then $Ad(g)$ is an isomorphism of $\mathfrak{a}(\mathfrak{g},p)$ onto $\mathfrak{a}(\mathfrak{g},q)$. This implies that $g^*dV_p=\pm dV_q$, hence dV is well-defined on M because M is orientable.

Since $\partial_i \in \mathfrak{a}(\Gamma_{dV_o}(T_B),0)$ for any $1 \leq i \leq n$, there is $u_1,\ldots,u_n \in \mathfrak{g}$ such that the Taylor expansion of u_i at the origin is given by $\partial_i + 0 + * + * + \cdots$. Since $\exp t_i u_i \in G$, $\varphi(t_1,\ldots,t_n) = (\exp t_1 u_1)\cdots(\exp t_n u_n)(p)$ is a smooth mapping of $R^n \ni (t_1,\ldots,t_n)$ into M and for sufficiently small neighborhood of $0 \in R^n$, φ gives a smooth coordinate at p. Consider $(\exp -t_n u_n)^* \cdots (\exp -t_1 u_1)^* dV_p$. This is obviously a smooth n-form on a neighborhood of p, but it is clear that this is equal to $dV_{\varphi(t_1,\ldots,t_n)}$. Thus, dV is a smooth n-form on M. Evidently, $G \subset \mathcal{D}_{dV}$ and hence $\mathfrak{g} \subset \Gamma_{dV}(T_M)$.

(3) The case $\mathfrak{a}(\mathfrak{g},p) \cong \mathfrak{a}(\Gamma_{\Omega_o}(T_B),0)$: We see $\mathfrak{a}(\mathfrak{g},(x_1,\ldots,x_{2m}),p) = \mathfrak{a}(\Gamma_{\Omega_o}(T_B),0)$ by a suitable choice of local coordinate. Define a 2-form $\Omega_p = \sum dx_i \wedge dx_{m+i}$. Since $\partial_i \in \mathfrak{a}(\Gamma_{\Omega_o}(T_B),0)$, we can take $u_i \in \mathfrak{g}$ as above and by the same reasoning, we have that Ω is a smooth 2-form such that $\Omega^m \neq 0$. Moreover, since $\exp t u_k \in G$, we have $\mathcal{L}_{u_k}\Omega = 0$, because $g^*\Omega = \Omega$ for any $g \in G$. Let $\Omega = \sum f_{ij} dx_i \wedge dx_j$. Then, we have $(\partial_k f_{ij})(p) dx_i \wedge dx_j = (\mathcal{L}_{u_k}\Omega)(p) = 0$. Thus, we have that Ω is a symplectic form and $G \subset \mathcal{D}_\Omega$, hence $\mathfrak{g} \subset \Gamma_\Omega(T_M)$.

(4) The case $\mathfrak{a}(\mathfrak{g},p) \cong \mathfrak{a}(\Gamma_{\omega_o}(T_B),0)$: We have $\mathfrak{a}(\mathfrak{g},(x_o,\ldots,x_{2m}),p) = \mathfrak{a}(\Gamma_{\omega_o}(T_B),0)$ by a suitable choice of local coordinate. Using this coordinate, we define a one dimensional subspace $L_p = \{\lambda dx_o : \lambda \in R\}$ in T_p^*M for any p. We see easily that $g^*L = L$ for any $g \in G$.

Now, obviously ∂_o, $\partial_i - x_{m+i}\partial_o$, ∂_{m+i}, $i = 1 \sim m$, belongs to $\Gamma_{\omega_o}(T_B)$. Thus, there are u_o, u_i, $u_{m+i} \in \mathfrak{g}$ such that the Taylor expansion of these at p are equal to

$$\partial_o + 0 + * + * + \cdots , \qquad \partial_i - x_{m+i}\partial_o + * + * + \cdots ,$$

$$\partial_{m+i} + 0 + * + * + \cdots$$

respectively.

Since $u_o(p), u_1(p),\ldots, u_{2m}(p)$ are linearly independent, we see by the same reasoning that L is a smooth subbundle of T^*_M.

Since M is orientable, L is a trivial bundle, hence there is a non-zero smooth section ω. Remark that $\mathcal{L}_{u_i}\omega = \tau\omega$. Thus, $(\mathcal{L}_{u_i}\omega)(p) = \lambda dx_o$, $\lambda \in R$. This implies that $\omega = f dx_o + \sum g_i dx_i + \sum (x_i + h_i) dx_{m+i}$ with $f(p) \neq 0$, $g_i(p) = h_i(p) = 0$ and $(dg_i)(p) = (dh_i)(p) = 0$. Hence, $(\omega \wedge (d\omega)^m)(p) \neq 0$. Thus, ω is a smooth contact form. Therefore, $G \subset \tilde{\mathfrak{D}}_\omega$ and $\mathfrak{J} \subset \Gamma_\omega(T_M)$.

9.2.4 Lemma <u>Let</u> V <u>be an open contractible neighborhood of</u> $p \in M$. <u>For any vector field</u> $u \in \Gamma_\alpha(T_V)$, <u>there is</u> $\tilde{u} \in \Gamma_\alpha(T_M)$ <u>such that</u> $u \equiv \tilde{u}$ <u>on a sufficiently small neighborhood of</u> p, <u>where</u> α <u>means one of</u> $\emptyset$, dV, Ω, ω, $\Gamma_\emptyset(T_M) = \Gamma(T_M)$.

Proof. The case $u \in \Gamma(T_V)$ is trivial, and the case $u \in \Gamma_\omega(T_V)$ is easy by using 8.3.1 Lemma. If $u \in \Gamma_\Omega(T_V)$, then $\Omega \lrcorner u$ is a closed 1-form on V, hence there is a function f such that $df = \Omega \lrcorner u$. Conversely, for any function f on V, $\Omega^{-1}(df)$ is contained in $\Gamma_\Omega(T_V)$. This proves the existence of $\tilde{u}$. If $u \in \Gamma_{dV}(T_V)$, then $dV \lrcorner u$ is a closed n-1-form on V. Thus, there is $\beta \in \Gamma(\Lambda_V^{n-2})$ such that $dV \lrcorner u = d\beta$. Conversely, for any $\beta \in \Gamma(\Lambda_V^{n-2})$ we can make a vector field $u \in \Gamma_{dV}(T_V)$. This proves the existence of $\tilde{u}$. (Cf. 8.5.2.)

Now, we are ready to prove 9.2.1 Theorem. For dV, Ω, ω we can take a local coordinate system such that $dV = dx_1 \wedge \cdots \wedge dx_n$, $\Omega = \sum dx_i \wedge dx_{m+i}$, $\omega = dx_o + \sum x_i dx_{m+i}$. Thus, by using 9.2.4 Lemma, we see that $\mathfrak{A}(\mathfrak{J}, p)$ is equal to one of $\mathfrak{A}(\Gamma(T_M), p)$, $\mathfrak{A}(\Gamma_{dV}(T_M), p)$, $\mathfrak{A}(\Gamma_\Omega(T_M), p)$, $\mathfrak{A}(\Gamma_\omega(T_M), p)$ for any $p \in M$. Now, we use the condition (c), and we see that $\mathfrak{J}$ is equal to one of $\Gamma(T_M)$, $\Gamma_{dV}(T_M)$, $\Gamma_\Omega(T_M)$ $\Gamma_\omega(T_M)$.

Assume at first that $\mathfrak{J}$ is equal to one of the above Lie algebras except $\Gamma_\omega(T_M)$, i.e. $\alpha = \emptyset$, dV, Ω. Then, the arguments in the sections VIII.1 ~ 2 show that there is a strong ILH-Lie subgroup G' of $\mathfrak{D}$ with the Lie algebra $\Gamma_\alpha(T_M)$. 7.1.5 Theorem shows that G' is the connected component of G under the LPSAC-topology. Moreover, by the arguments in VIII.1 ~ 2, we see that G' is the connected component of $\mathfrak{D}_\alpha$ under the relative topology. Therefore, every connected component of G under the

LPSAC-topology is equal to a connected component of $\mathcal{D}_\alpha$. Since $\mathcal{D}_\alpha$ satisfies the second countability axiom, the number of connected components are countable, hence G has a countable number of components under the LPSAC-topology. Therefore, G satisfies the second countability axiom under the LPSAC-topology. Recollect the arguments in pages VII.4 - 5. We have that G is a strong ILH-Lie subgroup of $\mathcal{D}$. Obviously, G is an open subgroup of $\mathcal{D}_\alpha$.

Now, let $\mathfrak{J} = \Gamma_\omega(T_M)$. Recall that $\Gamma_\omega(T_M) = \{ h\xi_\omega + d\omega^{-1}(dh - (dh \lrcorner \xi_\omega)\omega) : h \in \Gamma(1_M)\}$. We have a natural isomorphism φ of $\Gamma_\omega(T_M)$ onto Γ_ω. Moreover, we have a natural isomorphism Φ of $\tilde{\mathcal{D}}_\omega$ onto $\mathcal{D}_\omega$ defined by $\Phi(g) = (\tau, g)$, where $g^*\omega = \tau^{-1}\omega$. It is easy to see that $\Phi(c(t))$ is a piecewise C^1-curve in $\mathcal{D}_\omega$ if and only if $c(t)$ is a piecewise C^1-curve in $\tilde{\mathcal{D}}_\omega$. Since $G \subset \tilde{\mathcal{D}}_\omega$, we have $\Phi(G) \subset \mathcal{D}_\omega$ and the LPSAC-topologies for G and $\Phi(G)$ coincide, that is, $\Phi : G \mapsto \Phi(G)$ is bi-continuous under the LPSAC-topologies. Thus, by the same argument in VIII.3 and the argument as above, we see that $\Phi(G)$ is an open subgroup of $\mathcal{D}_\omega$. Therefore, the desired result can be obtained by defining the ILH-Lie group structure through Φ^{-1}, i.e. G is a strong ILH-Lie group (but not a strong ILH-Lie <u>subg</u>roup of $\mathcal{D}$) and an open subgroup of $\tilde{\mathcal{D}}_\omega$.

§ X Lie algebras of vector fields

X.1 Statement of the theorem and the idea of proof.

It is known in Pursell and Shanks [39] that an isomorphism between Lie algebras of infinitesimal automorphisms of C^∞-structures on manifolds M and N, that is, an isomorphism between C^∞-vector fields of M and N, yields an isomorphism between C^∞-structures of M and N.

The purpose of this chapter is to show that this is still true for some other structures on manifolds. In this chapter, manifolds are not assumed to be compact.

Let M and N be Hausdorff and finite dimensional manifolds without boundary.

The structures α which we are going to consider is one of the following :

(1) C^∞-structure, ($\alpha = \emptyset$)

(2) SL-structure, i.e. a volume element (positive n-form) with a non-zero constant multiplicative factor, ($\alpha = dV$)

(3) S_p- (symplectic) structure, i.e. symplectic 2-form with a non-zero constant multiplicative factor, ($\alpha = \Omega$)

(4) Contact structure, i.e. contact 1-form with a non zero C^∞-function as a multiplicative factor, ($\alpha = \omega$)

(5) Fibering with compact fibre, ($\alpha = \mathcal{F}$).

Let α (resp. α') be one of the above structures on M (resp. N). We denote by $\Gamma_\alpha(T_M)$ (resp. $\Gamma_{\alpha'}(T_N)$) the Lie algebra of all C^∞, α- (resp. α'-) preserving infinitesimal transformations with compact support and by $\mathcal{D}_\alpha(M)$ (resp. $\mathcal{D}_{\alpha'}(N)$) the group of all C^∞, α- (resp. α'-) preserving diffeomorphisms on M (resp. N) with compact support, that is, identity outside a compact subset.

Now, we can state the theorem.

Theorem *If $\Gamma_\alpha(T_M)$ is algebraically isomorphic to $\Gamma_{\alpha'}(T_N)$, then (M,α) is isomorphic to (N,α'). Especially, $\mathcal{D}_\alpha(M)$ is isomorphic to $\mathcal{D}_{\alpha'}(N)$.*

If α and α' are C^∞-structures, the above theorem is just the same as Pursell and Shanks. If α and α' are others, the precise meaning of the above statement should be the following : Under the same hypothesis of the above theorem, there is a C^∞-diffeomorphism φ of M onto N such that

(2') if dV, dV' are volume elements on M, N, then $\varphi^* dV' = C dV$, C = constant,

(3') if Ω, Ω' are symplectic forms on M, N, then $\varphi^*\Omega' = C\Omega$, C = constant,

(4') if ω, ω' are contact forms on M, N, then $\varphi^*\omega' = \tau\omega$, where τ is a non-zero C^∞-function,

(5') if $\mathfrak{F}$, $\mathfrak{F}'$ are fibering with compact fibres on M, N, then φ is a fibre preserving diffeomorphism.

The proof of our theorem is almost pararell to that of Pursell and Shanks. Essential ones are the first four cases. The last one is rather simple application of those .

First of all, we explain the idea for the cases of $\alpha = \emptyset$, dV, Ω, ω. At the begining we prove the following two lemmas which tell us how to choose standard coordinates :

Lemma 1 Suppose α is a smooth volume element (resp. a smooth symplectic form). If $u \in \Gamma_\alpha(T_M)$ does not vanish at $p \in M$, then there are a neighborhood U of p and a smooth coordinate $(x^1,\dots,x^n)$ on U such that $u \equiv \partial_1$ on U and $\alpha \equiv dx^1\wedge\dots\wedge dx^n$ (resp. $\alpha = \sum_{i=1}^{m} dx^i \wedge dx^{m+i}$, $n = 2m$) on U, where we use the notation ∂_i instead of $\frac{\partial}{\partial x^i}$.

Lemma 2 Suppose ω is a contact form. If $u \in \Gamma_\omega(T_M)$ does not vanish at $p \in M$, then there exist a neighborhood U of p, a smooth coordinate $(x^0,x^1,\dots,x^n)$, $n = 2m$, on U and a smooth function f on U such that $f\omega \equiv dx^0 + \sum_{i=1}^{m} x^{m+i} dx^i$ on U and $u \equiv \partial_0$ or ∂_1 on U.

These two lemmas make all computations very simple and by appropriate choice of elements and computations of Lie brackets, we can prove the following :

Proposition 1 <u>For any maximal ideal $\mathcal{I}$ of $\Gamma_\alpha(T_M)$ which does not conatin $[\Gamma_\alpha(T_M),\Gamma_\alpha(T_M)]$, there is a unique point $p \in M$ such that $\mathcal{I} = \mathcal{I}_p$, where $\mathcal{I}_p$ is the ideal consisting of all element u such that u and its all derivatives vanish at the point p.</u>

Once we know this, we can use a rather standard method to prove the theorem. This is in fact the method of Pursell and Shanks. However, at the first stage, we should remark the following, though it is not directly relevant to our purpose.

Proposition 2 $[\Gamma_\alpha(T_M),\Gamma_\alpha(T_M)] = \Gamma_\alpha(T_M)$, <u>if α is a C^∞-structure or a contact structure. Moreover, $\Gamma_{dV}(T_M)/[\Gamma_{dV}(T_M),\Gamma_{dV}(T_M)] = H^{n-1}(M)$ and $\Gamma_\Omega(T_M)/[\Gamma_\Omega(T_M),\Gamma_\Omega(T_M)] = H^1(M)$, where $n = \dim M$ and $H^*(M)$ is the de Rham cohomology group of compact support.</u>

Anyhow, by Proposition 1, we see that if $\Phi : \Gamma_\alpha(T_M) \mapsto \Gamma_{\alpha'}(T_N)$ is an isomorphism, then there is a mapping $\varphi : M \mapsto N$ such that $\Phi(\mathcal{I}_p) = \mathcal{I}_{\varphi(p)}$. This is obviously one to one and surjective. Remark that for any subset A in M, the closure $\bar{A}$ is equal to $\{ q \in M : \mathcal{I}_q \supset \cap\{\mathcal{I}_p : p \in A\}\}$. Since this relation is transfared by Φ, we see that $\varphi : M \mapsto N$ is a homeomorphism.

To prove φ is an isomorphism of (M,α) onto (N,α'), we need to characterize the zeros of an element $u \in \Gamma_\alpha(T_M)$.

Lemma 3 <u>For any</u> $u \in \Gamma_\alpha(T_M)$, $u(p) \neq 0$ <u>if and only if</u> $\Gamma_\alpha(T_M) = [u,\Gamma_\alpha(T_M)] + \mathcal{I}_p$.

Therefore, if $u_1,\ldots,u_k \in \Gamma_\alpha(T_M)$ are linearly independent at p, then $\Phi(u_1), \ldots,\Phi(u_k)$ are also linearly independent at $\varphi(p)$. Moreover, if $\sum f_i u_i \in \Gamma_\alpha(T_M)$, then $\Phi(\sum f_i u_i)(q) = \sum f_i(\varphi^{-1}(q))\Phi(u_i)(q)$. So, by a suitable choice of $u_1,\ldots,u_k$ and $f_1,\ldots,f_k$, the above fact shows that φ is a C^∞-diffeomorphism and keeps the structures.

The case $\alpha = \mathcal{F}$: If this is the case, a maximal ideal does not correspond to a point on M but a point of the base space B_M of the fibering of M. We denote by $\Gamma_F(T_M)$ the intersection of all maxiamal ideals of $\Gamma_{\mathfrak{F}}(T_M)$. $\Gamma_F(T_M)$ is an ideal of $\Gamma_{\mathfrak{F}}(T_M)$ consisting of all infinitesimal transformations such that every fibre is left fixed. So, we consider an ideal $\mathcal{J}$ of $\Gamma_{\mathfrak{F}}(T_M)$ such that $\mathcal{J} \subset \Gamma_F(T_M)$ and maximal in $\Gamma_F(T_M)$. We say such an ideal a maximal ideal of $\Gamma_{\mathfrak{F}}(T_M)$ in $\Gamma_F(T_M)$. Then, this corresponds to a point of M, and permit us to repeat the same procedure as above.

Finally, we shall remark that in this case, the Lie algebra $\Gamma_{\mathfrak{F}}(T_M)$ is not primitive anymore, and the above method of the proof can be applied to various non-primitive Lie algebras. So, it seems to be natural to conjecture that there is a huge class of structures on manifolds on which the theorem holds.

However, the argument used here can not be applied for regular contact structures. $\Gamma_{s\omega}(T_M)$ contains one dimensional center $R\xi_\omega$ and this is equal to the intersection of all maximal ideals of $\Gamma_{s\omega}(T_M)$. This is too small. We can not capture a point of an orbit of S^1. (But we can capture a point in the orbit space.) Therefore, an example may come from regular contact structures where the theorem does not hold.

X.2 Proof of Lemmas 1, 2.

These are proved case by case.

(i) The case $\alpha = dV$.

Let $u \in \Gamma_{dV}(T_M)$ such that $u(p) \neq 0$ at $p \in M$. Then, there is a coordinate $x^1,\dots,x^n$ at p such that $u \equiv \partial_1$ on a neighborhood of p. The volume element dV is described by $f dx^1 \wedge \cdots \wedge dx^n$. However, since $u \in \Gamma_{dV}(T_M)$, we see that the Lie derivative $\mathcal{L}_{\partial_1} dV = \partial_1 f\, dx^1 \wedge \cdots \wedge dx^n = 0$, hence f does not contain the variable x^1. $f dx^2 \wedge \cdots \wedge dx^n$ is a volume element defined on a neighborhood of the origin of R^{n-1} and hence by a suitable choice of coordinate, this is written by $d\acute{x}^2 \wedge \cdots \wedge d\acute{x}^n$. Thus, $x^1, \acute{x}^2, \dots, \acute{x}^n$ is the desired coordinate.

(ii) The case $\alpha = \Omega$.

Let $u \in \Gamma_\Omega(T_M)$ such that $u(p) \neq 0$ at $p \in M$. Then, there is a coordinate $x^1,\dots,x^n$ at p, $n = 2m$, such that $u \equiv \partial_1$ on a neighborhood of p. Since $\mathcal{L}_{\partial_1}\Omega \equiv 0$ on a neighborhood of p, the coefficients of $\Omega \equiv \sum \Omega_{ij} dx^i \wedge dx^j$ does not contain the variable x^1. We put $\Omega = \Omega_1 + \Omega_2$, $\Omega_1 = dx^1 \wedge \sum_{j=2}^{n} \Omega_{1j}\, dx^j$, $\Omega_2 = \sum_{2 \leq i \leq j} \Omega_{ij} dx^i \wedge dx^j$. Since $d\Omega = 0$ and Ω is of rank $2m$, we see that $d(\sum_{j=2}^{n} \Omega_{1j} dx^j) \equiv 0$, and Ω_2 is a closed form and of rank $2(m-1)$ on a neighborhood of the origin of R^{n-1}. Therefore, there are a C^∞-function f and a C^∞- 1-form θ on a neighborhood of the origin of R^{n-1} such that $\Omega_1 = dx^1 \wedge df$ and $\Omega_2 = d\theta$.

Now, by adding dg if necessary, we may assume that $\theta(p) \neq 0$. Thus, $\theta \wedge (d\theta)^{m-1} \neq 0$. (Cf. [1],p133.) Hence by Darboux's theorem, there is a coordinate $x'^2,\dots,x'^n$ on a neighborhood of the origin of R^{n-1} such that $\theta = dx'^{m+1} + \sum_{i=2}^{m} x'^{m+i} dx'^i$. Thus, we have $\Omega = dx^1 \wedge df + \sum_{i=2}^{m} dx'^{m+i} \wedge dx'^i$.

Since $\Omega^m \neq 0$, we see that $\partial'_{m+1} f \neq 0$. Thus the desired coordinate is obtained by the change of coordinate $(x^1, x'^i, x'^{m+1}, x'^{m+i}) \longmapsto (x^1, x'^{m+i}, f, x'^i)$, $i > 1$.

(iii) The case $\alpha = \omega$.

Let $u \in \Gamma_\omega(T_M)$ such that $u(p) \neq 0$. Then, there is a coordinate $x^o, x^1, \cdots, x^m, x^{m+1}, \cdots, x^n$ at p such that $u \equiv \partial_o$ on the coordinate neighborhood. We put $\omega = \sum_{i=o}^{n} \omega_i\, dx^i$. Since $\mathcal{L}_{\partial_o}\omega = \tau\omega$, we have $\partial_o \omega_i = \tau\omega_i$, $0 \leq i \leq n$. Hence

$$\omega_i = \omega_i(0, x^1, \dots, x^n) \exp \int \tau\, dx^o.$$

Thus, there is a C^∞-function F on a neighborhood of p such that $\omega' = F\omega$ has coefficients ω'_i which does not contain the variable x^o.

(iii,a) Suppose at first that $\omega'_o(p) \neq 0$. Then, on a neighborhoof of p, we have $\omega' = \omega'_o(dx^o + \sum_{i=1}^{n} \omega''_i\, dx^i)$. Thus, we may put $\omega' = dx^o + \sum \omega'_i dx^i$. Now, $d\omega' = d\omega'_i \wedge dx^i$ is a symplectic form on a neighborhood of the origin of $R^n = R^{2m}$, hence by a suitable choice of coordinate $d\omega'$ is described by $\sum d\acute{x}^{m+i}\wedge d\acute{x}^{i}$. Therefore, $\omega' = dx^{o} + df + \sum \acute{x}^{m+i} d\acute{x}^{i}$ for a C^{∞}-function f such that $f(p) = 0$. Put $\acute{x}^{o} = x^{o} + f$. Since f does not contain the variable x^{o}, we see easily that $\partial_{o} \equiv \partial'_{o}$ where ∂'_{o} means $\frac{\partial}{\partial \acute{x}^{o}}$. Thus, $\acute{x}^{o},\ldots,\acute{x}^{n}$ is the desired coordinate.

(iii,b) Suppose $\omega'_{o}(p) = 0$. Then, the characteristic vector field $\xi_{\omega'}$ has coefficients which does not contain the variable x^{o}, hence $[\,\partial_{o}, \xi_{\omega'}] = 0$. Thus, ther is a coordinate $\acute{x}^{1},\ldots,\acute{x}^{n}$ on a neighborhood of the origin of R^{n-1} such that $\xi_{\omega'} \equiv \partial'_{1}$. Now, interchange the numbering of the coordinate x^{o} and $\acute{x}^{1}$, and we have $u \equiv \partial_{1}$, $\xi_{\omega'} \equiv \partial'_{o}$. ω' must be expressed by $d\acute{x}^{o} + \sum_{i=1}^{n} \omega''_{i} d\acute{x}^{i}$. Again, ω''_{i} does not contain the variable x^{1}. Moreover, since $\mathcal{L}_{\xi_{\omega'}} \omega' = 0$, we see that ω''_{i} does not contain the variable $\acute{x}^{o}$. Thus, $d\omega'$ can be regarded as a symplectic 2-form on a neighborhood of the origin of $R^{n} = R^{2m}$ and $\mathcal{L}_{\partial_{1}} d\omega' = 0$.

Thus, we have the same situation as the case $\alpha = \Omega$. Therefore, there is a coordinate $\grave{x}^{2},\ldots,\grave{x}^{n}$ such that $d\omega' = d\grave{x}^{m+1}\wedge dx^{1} + \sum_{i=2}^{m} d\grave{x}^{m+i}\wedge d\grave{x}^{i}$, hence $\omega' = d\acute{x}^{o} + df + \grave{x}^{m+1}dx^{1} + \sum_{i=2}^{m} \grave{x}^{m+i} d\grave{x}^{i}$, where $f = f(x^{1},\grave{x}^{2},\ldots,\grave{x}^{n})$. Put $\grave{x}^{o} = \acute{x}^{o} + f$. The desired coordinate is given by $\grave{x}^{o},x^{1},\grave{x}^{2},\ldots,\grave{x}^{n}$.

X.3 Lemmas 1, 2 $\Rightarrow$ Proposition 1.

The goal here is to prove Proposition 1.

Let u_{o} be an element in $\Gamma_{\alpha}(T_{M})$ such that $u_{o}(p) \neq 0$ at $p \in M$. Then, by Lemmas 1,2, there is a coordinate neighborhood U of p with a coordinate $x^{1},\ldots, x^{n}$ or $x^{o},x^{1},\ldots,x^{m},x^{m+1},\ldots,x^{n}$ such that $u_{o} \equiv \partial_{1}$ or ∂_{o} on U and α has a standard expression as in Lemmas 1,2. Here, we always assume that $u_{o} \equiv \partial_{1}$ by an appropriate change of the numbering of coordinate. Let V be an open ball in that coordinate neighborhood U with the center at p = the origin such that $\overline{V} \subset U$.

10.3.1 Lemma <u>For any</u> $v \in \Gamma_{\alpha}(T_{M})$ <u>such that the support of</u> v <u>is contained in</u> V, <u>there is</u> $u \in \Gamma_{\alpha}(T_{M})$ <u>such that</u> support $u \subset U$ <u>and</u> $[\,\partial_{1}, u\,] \equiv v$ <u>on</u> V.

Proof. Denote $I_1(v) = \sum_{i=1}^{n}\int_{-\infty}^{x^1} v_i(t,x^2,\dots,x^n)dt\ \partial_i$. Then, easily $[\partial_i, I_1(v)] = I_1([\partial_i, v])$ for any i. Therefore, we have

$$\operatorname{div} I_1(v) = \int_{-\infty}^{x^1} \operatorname{div} v\, dx^1 = 0$$

$$d(\Omega \lrcorner I_1(v)) = \int_{-\infty}^{x^1} (\Omega \lrcorner v)dx^1 = 0$$

$$d(\omega \lrcorner I_1(v)) + d\omega \lrcorner I_1(v) = \int_{-\infty}^{x^1} (d(\omega \lrcorner v) + d\omega \lrcorner v)dx^1 = 0.$$

Now, remark that $I_1(v)$ can be defined on a neighborhood V' of $\bar{V}$ and $I_1(v) \in \Gamma_\alpha(T_{V'})$. Thus, by the same manner as in 9.2.4 Lemma, there is u such that support $u \subset U$ and $u \equiv I_1(v)$ on V. Obviously, $[\partial_1, I_1(v)] \equiv v$ on V.

10.3.2 Lemma <u>Suppose $\mathcal{J}$ is an ideal of $\Gamma_\alpha(T_M)$ such that for any point $p \in M$, there is $u \in \mathcal{J}$ such that $u(p) \neq 0$. Then, $\mathcal{J}$ contains $[\Gamma_\alpha(T_M),\Gamma_\alpha(T_M)]$.</u>

Proof. For any point $p \in M$, we take neighborhoods U_p, V_p with the same property as the previous lemma, and by an appropriate change of the numbering of coordinate, we may assume that for any U_p, there is $u_p \in \mathcal{J}$ such that $u_p \equiv \partial_1$ on U_p.

Now, for the proof, it is enough to show that $\mathcal{J}$ contains $[u,v]$ for any $u, v \in \Gamma_\alpha(T_M)$ such that support u and support v are contained in V_p. By 10.3.1, we see that a suitable extension of $I_1(v)$ is contained in $\Gamma_\alpha(T_M)$. So, take the identity

$$[\partial_1, [u, I_1(v)]] = [[\partial_1, u], I(v)] + [u, [\partial_1, I_1(v)]].$$

Since $\partial_1 \in \mathcal{J}$ and support $u \subset V_p$, we see that the left hand side and similarly the first term of the right hand side are in $\mathcal{J}$. Since the last term is equal to $[u,v]$ on V_p, we have the desired result.

Now, we are ready to prove Proposition 1. Suppose $\mathcal{J}$ is a maximal ideal of $\Gamma_\alpha(T_M)$ which does not contain $[\Gamma_\alpha(T_M),\Gamma_\alpha(T_M)]$. By 10.3.2, there must be a point $p \in M$ such that all elements of $\mathcal{J}$ vanish at p.

Now, assume that there is $v \in \mathcal{J}$ such that some of the derivatives of v does not vanish at p. Then, by taking $[\partial_i, v]$ appropriately and successively, we can conclude easily that $\mathcal{J}$ must contain an element which does not vanish at p. Therefore, we have that all derivative of $v \in \mathcal{J}$ must vanish at p, i.e. $\mathcal{J} \subset \mathcal{J}_p$. However, the maximality of $\mathcal{J}$ implies $\mathcal{J} = \mathcal{J}_p$.

X.4 Proof of Proposition 2.

This will be done case by case.

(i) The case $\alpha = C^\infty$-structure.

It is enough to show that any $u \in \Gamma(T_M)$ with a small support is contained in $[\Gamma(T_M), \Gamma(T_M)]$. So, we may assume $u = \sum_{i=1}^{n} u_i \partial_i$. Again, it is enough to prove that $g\,\partial_i$ such that the support of g is very small is contained in $[\Gamma(T_M), \Gamma(T_M)]$. So, take the identity

$$[g\,\partial_i\,,\, x^i \partial_i] + [\partial_i\,,\, x^i g\,\partial_i] = 2g\,\partial_i\,.$$

The desired result follows immediately.

(ii) The case α = contact structure.

Consider the one to one correspondence between $\Gamma_\omega(T_M)$ and $\Gamma(1_M)$ in 8.3.1. By this correspondence, the Lie bracket is transfered to the following : (See [13].)

$$\{f,g\} = \sum_{i=1}^{m} \partial_{m+i} f\,(\,\partial_i g + x^{m+i}\partial_o\,) - \sum_{i=1}^{m} \partial_{m+i}(\,\partial_i + x^{m+i}\partial_o\,) + (\partial_o f)g - (\partial_o g)f.$$

Thus, by a simple computation, we have

$$\sum_{i=1}^{m} \{x^{m+i} f, x^i\} - \sum_{i=1}^{m} \{x^i x^{m+i} f, 1\} - \{f, x^o\} + \{x^o f, 1\} = (m+2)f.$$

Remark that it is enough to show that any $f \in \Gamma(1_M)$ with a small support is contained in $\{\Gamma(1_M), \Gamma(1_M)\}$. The above equality shows the desired result.

(iii) The cases $\alpha = dV$ and ω.

Recall the formulae $dV \lrcorner [u,v] = d(dV \lrcorner u \lrcorner v)$, $\Omega \lrcorner [u,v] = d(\Omega \lrcorner u \lrcorner v)$. Then, we see $\Gamma_\delta(T_M) \supset [\Gamma_{dV}(T_M), \Gamma_{dV}(T_M)]$ and $\Gamma_\partial(T_M) \supset [\Gamma_\Omega(T_M), \Gamma_\Omega(T_M)]$. It is enough to prove the equalities.

To prove these, we need the following :

10.4.1 Lemma <u>Let</u> $x^1,\dots,x^n$ <u>be a local coordinate such that</u> $dV = dx^1\wedge\cdots\wedge dx^n$ <u>or</u> $\Omega = \sum_{i=1}^{m} dx^i\wedge dx^{m+i}$. <u>Suppose</u> $u \in \Gamma_\alpha(T_M)$, $\alpha = dV$ <u>or</u> Ω, <u>has the support contained in an open cube</u> V <u>of the coordinate neighborhood. If</u> $\int_V u dx^1\cdots dx^n = 0$, <u>then</u> u <u>is contained in</u> $[\Gamma_\alpha(T_M),\Gamma_\alpha(T_M)]$.

Proof. We may assume that V is a 2ε-cube with the center at the origin. Let $\phi(t)$ be a C^∞-function on R such that $\int_{-\infty}^{\infty}\phi(t)dt = 1$ and $\phi = 0$ for $t \notin (-\varepsilon,\varepsilon)$. Put $\bar{u}_{(1,\dots,k)} = \int_{-\varepsilon}^{\varepsilon}\cdots\int_{-\varepsilon}^{\varepsilon} u dx^1\cdots dx^k$. Remark that ∂_i, $1 \leq i \leq n$, is contained in $\Gamma_\alpha(T_V)$ and hence $[\partial_i, v] \in [\Gamma_\alpha(T_M),\Gamma_\alpha(T_M)]$ for any $v \in \Gamma_\alpha(T_M)$ such that support $v \subset V$. Recall the definition of $I_1(v)$ in the proof of 10.3.1. We define similarly $I_i(v) = \int_{-\infty}^{x^i} v(x^1,\dots,t,\dots,x^n)dt$.

Now, since support $u \subset V$, we see support $I_1(u - \phi(x^1)\bar{u}_{(1)}) \subset V$ and

$$[\partial_1, I_1(u - \phi(x^1)\bar{u}_{(1)})] = u - \phi(x^1)\bar{u}_{(1)} .$$

Similarly, $I_k(\phi(x^1)\cdots\phi(x^{k-1})(\bar{u}_{(1,\dots,k-1)} - \phi(x^k)\bar{u}_{(1,\dots,k)}))$ has the support contained in V and the Lie bracket product with ∂_k is equal to

$$\phi(x^1)\cdots\phi(x^{k-1})(\bar{u}_{(1,\dots,k-1)} - \phi(x^k)\bar{u}_{(1,\dots,k)}).$$

Since $\bar{u}_{(1,\dots,n)} = 0$, by summing up these equalities, we see that $u \in [\Gamma_\alpha(T_M),\Gamma_\alpha(T_M)]$.

Now, let $u \in \Gamma_\delta(T_M)$ (resp. $\Gamma_\partial(T_M)$). Then, there is $n-2$- (resp. 0-) form γ such that $dV \lrcorner u = d\gamma$ (resp. $\Omega \lrcorner u = d\gamma$). For the proof of Proposition 2, it is enough to show that $u \in [\Gamma_\alpha(T_M),\Gamma_\alpha(T_M)]$, if γ has the support contained in an open cube in a standard coordinate neighborhood V. So, by the previous lemma, we have only to show that $\int_V u\, dx^1\cdots dx^n = 0$.

(a) The case $u \in \Gamma_\delta(T_M)$.

Let $\gamma = \sum_{i<j} \gamma_{ij}\, dx^1\wedge \ldots\wedge \check{dx}^i\wedge \ldots\wedge \check{dx}^j\wedge\ldots \wedge dx^n$. Though γ_{ij} are given for the pairs (i,j) such that $i < j$, we extended these by putting $\gamma_{ij} = -\gamma_{ji}$, so that γ_{ij} are given for all pairs (i,j). Now, let $v_j = \sum_{i=1}^{n}(-1)^i\gamma_{ij}\,\partial_i$. Then, by a simple computation, we see that

$$d\gamma = \sum_{j=1}^{n} (\operatorname{div} v_j)\, dx^1\wedge\ldots\wedge \check{dx}^j\wedge\ldots\wedge dx^n .$$

Hence, $u = \sum_{j=1}^{n} (-1)^j(\operatorname{div} v_j)\,\partial_j$. Therefore $\int_V u\, dx^1\cdots dx^n = 0$ because support $v_j \subset V$.

(b) The case $u \in \Gamma_\partial(T_M)$.

We have $\Omega \lrcorner u = d\gamma$ and γ is a C^∞-function with support $\gamma \subset V$. Since Ω is non-degenerate, we have only to show $\Omega \lrcorner \int_V u\, dx^1\cdots dx^n = 0$. However

$$\Omega \lrcorner \int_V u\, dx^1\cdots dx^n = \int_V d\gamma\, dx^1\cdots dx^n = 0,$$

because Ω has constant coefficients in the standard coordinate.

X.5 Proof of Lemma 3.

Recall the statement of Lemma 3 in §X.1. If $u(p) \neq 0$, then any element $v \in \Gamma_\alpha(T_M)$ can be expressed by $[u, I_1(v)]$ near the point p. (Cf.10.3.1.) Thus, $v - [u, I_1(v)] \in \mathcal{J}_p$.

Conversely, if $[u, \Gamma_\alpha(T_M)] + \mathcal{J}_p = \Gamma_\alpha(T_M)$, then the 1-jet of u at p can not vanish, because otherwise all elements of $[u, \Gamma_\alpha(T_M)] + \mathcal{J}_p$ vanish at p. Now, assume $u(p) = 0$. Let $\Gamma_\alpha(T_M, p)$ be the isotropy subalgebra of $\Gamma_\alpha(T_M)$ at p, i.e. the subalgebra of all elements in $\Gamma_\alpha(T_M)$ which vanish at p. For any $v \in \Gamma_\alpha(T_M,p)$ we denote by $\bar{v}$ the linear part of v at p and by $\mathfrak{g}$ the subalgebra of $\mathfrak{gl}(n)$ consisting of all linear part $\bar{v}$ of $v \in \Gamma_\alpha(T_M,p)$.

Now, the equality $[u,\Gamma_\alpha(T_M)] + \mathcal{J}_p = \Gamma_\alpha(T_M)$ implies $[\bar{u}, \mathfrak{g}] = \mathfrak{g}$ because $u \in \Gamma_\alpha(T_M,p)$. But this is impossible because the linear mapping $ad(\bar{u}) : \mathfrak{g} \to \mathfrak{g}$ has a kernel $\bar{u}$.

X.6 Proof of Theorem for the cases $\alpha = \emptyset$, dV, Ω, ω.

Now, we can start to prove the theorem. Let $\Phi : \Gamma_\alpha(T_M) \to \Gamma_{\alpha'}(T_N)$ be an isomorphism. For any maximal ideal $\mathcal{J}_x$ of $\Gamma_\alpha(T_M)$, $\Phi(\mathcal{J}_x)$ is also a maximal ideal of $\Gamma_{\alpha'}(T_N)$. So, by Proposition 1, there is $y \in N$ such that $\Phi(\mathcal{J}_x) = \mathcal{J}_y$. We put $\varphi(x) = y$. This is obviously a one to one correspondence and in fact a homeomorphism by the reasoning in §X.1.

Remark that Lemma 3 shows immediately the following :

10.6.1 Lemma <u>If</u> $u_1,\dots,u_k \in \Gamma_\alpha(T_M)$ <u>are linearly independent on an open set</u> U, <u>then</u> $\Phi(u_1),\dots,\Phi(u_k)$ <u>are also linearly independent on the open set</u> $\varphi(U)$. <u>Moreover, if</u> $\sum_{i=1}^{k} f_i u_i \in \Gamma_\alpha(T_M)$, <u>then</u> $\Phi(\sum_{i=1}^{k} f_i u_i) = \sum_{i=1}^{k} (\varphi^{-1*}f_i)\Phi(u_i)$, <u>hence all</u> $\varphi^{-1*}f_i$ <u>are</u> C^∞ <u>function on</u> $\varphi(U)$.

If $\alpha = C^\infty$-structure, the above lemma shows that φ is a C^∞- diffeomorphism. If α are others, the theorem will be proved case by case. Recall what we have to prove is that φ is an α- preserving diffeomorphism.

(i) The case $\alpha = dV$.

For any $p \in M$, we choose a standard coordinate $x^1,\dots,x^n$. Recall the statement of 9.2.4. Clearly, $\Gamma_{dV}(T_M)$ contains suitable extensions of $\partial_1,\dots,\partial_n$. Since the argument is local, we use the same notation ∂_i for the extension of ∂_i. Now, let $v_i = \Phi(\partial_i)$. Then, $v_1,\dots,v_n$ are linearly independent and $[v_i,v_j] = 0$ on a neighborhood of $\varphi(p)$. Therefore, there is a coordinate $y^1,\dots,y^n$ such that $v_i = \partial'_i$ on a neighborhood of $\varphi(p)$, where $\partial'_i = \frac{\partial}{\partial y i}$.

Remark that $\Gamma_{dV}(T_M)$ contains a suitable extension of $x^i\partial_j$, $i \neq j$. (Again, we use the same notation for the extended vector field.) By the above lemma, we see $\Phi(x^i\partial_j) = f_{ij}\partial'_j$ with a smooth function f_{ij} and $f_{ij}(\varphi(q)) = x^i(q)$. Since $[\partial_k, x^i\partial_j] = \delta^i_k\,\partial_j$, we see that $\partial'_k f_{ij} = \delta^i_k$ and hence $\Phi(x^i\partial_j) = y^i\partial'_j$, $y^i(\varphi(q)) = x^i(q)$. This implies that φ is a smooth diffeomorphism and moreover, since $\partial'_i \in \Gamma_{dV'}(T_N)$, dV' must be $Cdy^1\wedge\ldots\wedge dy^n$, where C is a constant.

(ii) The case $\alpha = \Omega$.

For any $p \in M$, we choose a standard coordinate $x^1,\ldots,x^n$ such that $\Omega = \sum_{i=1}^{m} dx^i\wedge dx^{m+i}$. By 9.2.4 Lemma, we see that $\Gamma_\Omega(T_M)$ contains suitable extensions of $\partial_1,\ldots,\partial_n$. We put $v_i = \Phi(\partial_i)$. (Here, we use the same notation for the extended vector fields.) By the same reasoning as above, there is a smooth coordinate $y^1,\ldots, y^n$ such that $v_i = \partial'_i$ on a neighborhood of $\varphi(p)$.

Remark that $\Gamma_\Omega(T_M)$ contains suitable extensions of $x^i\partial_{m+i}$, $x^{m+i}\partial_i$ and $x^i\partial_j - x^{m+j}\partial_{m+i}$. Therefore, we have $y^i(\varphi(*)) = x^i(*)$ for $1 \leq i \leq n$ and suitable extensions of $y^i\partial'_{m+i}$, $y^{m+i}\partial'_i$, $y^i\partial'_j - y^{m+j}\partial'_{m+i}$ are contained in $\Gamma_{\Omega'}(T_N)$ together with ∂'_i, ∂'_{m+i}.

Now, since $\mathcal{L}_{\partial'_i}\Omega' = 0$, $0 \leq i \leq n$, on a neighborhood of $\varphi(p)$, we see that Ω' has the constant coefficients Ω'_{ij} on the neighborhood of $\varphi(p)$. However, since $y^i\partial'_{m+i}$, $y^{m+i}\partial_i$ are in $\Gamma_{\Omega'}(T_N)$, Ω' must be $\Omega' = \sum_{i=1}^{m}\Omega'_i\, dy^i\wedge dy^{m+i}$ and finally the fact that $y^i\partial'_j - y^{m+j}\partial'_{m+i} \in \Gamma_{\Omega'}(T_N)$ shows that $\Omega' = C'\sum_{i=1}^{m} dy^i\wedge dy^{m+i}$ on a neighborhood of $\varphi(p)$. This implies that φ is a diffeomorphism such that $\varphi^*\Omega' = C\Omega$.

(iii) The case of contact structures.

For any point $p \in M$, we take a standard coordinate $x^o, x^1,\ldots,x^n$, ($n = 2m$), i.e. $\omega = dx^o + \sum_{i=1}^{m} x^{m+i}dx^i$. Notice that $\Gamma_\omega(T_M)$ contains suitable extensions of ∂_o, ∂_i, $1 \leq i \leq m$, $\partial_{m+i} - x^i\partial_o$, $1 \leq i \leq m$. We use the same notations for the extended vector fields because all arguments here are local. We put $v_i = \Phi(\partial_i)$, $0 \leq i \leq m$, and $v_{m+i} = \Phi(\partial_{m+i} - x^i\partial_o)$. Then, $[v_i, v_j] = [v_{m+i}, v_{m+j}] = 0$ and $[v_i, v_{m+j}] = -\delta_{ij}v_o$ on a neighborhood of $\varphi(p)$. Therefore, there is a local coordinate $y^o, \ldots, y^m, y^{m+1}, \ldots, y^n$ at $\varphi(p)$ such that $\partial_i' = v_i$ for $0 \leq i \leq m$. Easily, we see $[\partial_i', v_{m+j} + y_j\partial_o'] = 0$ on a neighborhood of $\varphi(p)$. Thus, we can take a local coordinate $y^o, y^1, \ldots, y^m, y^{m+1}, \ldots, y^n$ such that $\partial'_{m+j} = v_{m+j} + y^j\partial_o'$.

Remark that $\Gamma_\omega(T_M)$ contains suitable extensions of the following vector fields :

$$x^{m+i}\partial_i - \frac{1}{2}(x^{m+i})^2\partial_o, \qquad x^i(\partial_{m+i} - x^i\partial_o) + \frac{1}{2}(x^i)^2\partial_o, \qquad x^o\partial_o + \sum_{i=1}^m x^i\partial_i,$$

$$(x^o + \sum_{i=1}^m x^i)\,\partial_o + \sum_{i=1}^m x^{m+i}(\partial_{m+i} - x^i\partial_o).$$

We put

$$\Phi(x^{m+i}\partial_i - \frac{1}{2}(x^{m+i})^2\partial_o) = g^i v_i - \frac{1}{2}(g^i)^2 v_o,$$

$$\Phi(x^i(\partial_{m+i} - x^i\partial_o) + \frac{1}{2}(x^i)^2\partial_o) = f^i\, v_{m+i} + \frac{1}{2}(f^i)^2\, v_o,$$

$$\Phi(\sum_{i=1}^m x^i\,\partial_i) = \sum_{i=o}^m f^i v_i,$$

$$\Phi((\sum_{i=o}^m x^i)\,\partial_o + \sum_{i=1}^m x^{m+i}(\partial_{m+i} - x^i\partial_o)) = (\sum_{i=o}^m f^i)\, v_o.$$

Here, we use the above 10.6.1, and moreover since $v_o, \ldots, v_n$ are linearly independent on a neighborhood at $\varphi(p)$, we see that $f^o, f^1, \ldots, f^m, g^1, \ldots, g^m$ are all C^∞-functions.

Remark that

$$[v_k, g^i v_i - \frac{1}{2}(g^i)^2 v_o] = \begin{cases} v_i & (k = m+i) \\ 0 & (\text{others}), \end{cases} \qquad [v_k, f^i v_{m+i} + \frac{1}{2}(f^i)^2 v_o] = \begin{cases} v_{m+i} & (k = i) \\ 0 & (\text{others}), \end{cases}$$

$$[v_k, f^o v_o + \sum_{i=1}^m f^i v_i] = v_k \quad \text{and} \quad [v_o, (\sum_{i=o}^m f^i)v_o + g^i v_{m+i}] = v_o.$$

The last two inequalities show that $f^1, \ldots, f^m, g^1, \ldots, g^m$ do not contain the variable y^o. The second equality gives us $f^i = y^i$ and the first together with $\partial_o g^i = 0$

gives us $g^i = y^{m+i}$, hence from the third, $f^o = y^o$. This implies $y^i(\varphi(*)) = x^i(*)$, hence φ is a C^∞-diffeomorphism and $\varphi^*(dy^o + \sum_{i=1}^{m} y^{m+i}dy^i) = \omega$.

Now, it remains to prove that $\omega' = \tau(dy^o + \sum_{i=1}^{m} y^{m+i}dy^i)$. Since suitable extensions of $\partial'_o, \ldots, \partial'_m$ are in $\Gamma_{\omega'}(T_N)$, we see that $\mathcal{L}_{\partial'_k}\omega' = \tau_k\omega'$, $0 \leq k \leq m$. This implies $\partial'_k\omega'_a = \tau_k\omega'_a$, $0 \leq a \leq n$, where $\omega' = \sum_{a=o}^{n} \omega'_a dy^a$. Thus, $\omega'_a = (\exp \int \tau_k dy^k)\omega''_a$ and ω''_a does not contain the variable y^k. Therefore, we find a non-zero C^∞-function τ defined locally and such that $\tau\omega' = \sum_{a=o}^{n} \bar{\omega}_a dy^a$ does not contain the variables $y^o, y^1, \ldots, y^m$. We put $\bar{\omega} = \sum_{a=o}^{n} \bar{\omega}_a dy^a$. Notice that $\Gamma_{\omega'}(T_N)$ contains a suitable extension of $\sum_{k=o}^{m} y^k \partial'_k$. Therefore, we have

$$\mathcal{L}_{\sum_{k=o}^{m} y^k \partial'_k} \bar{\omega} = \bar{\tau}\bar{\omega}$$

On the other hand,

$$\mathcal{L}_{\sum_{k=o}^{m} y^k \partial'_k} \bar{\omega} = d(\bar{\omega} \lrcorner \sum_{k=o}^{m} y^k \partial'_k) + d\bar{\omega} \lrcorner \sum_{k=o}^{m} y^k \partial'_k$$

$$= \sum_{k=o}^{m} \bar{\omega}_k dy^k + \sum_{k=o}^{m} \sum_{j=m+1}^{n} y^k \bar{\omega}_{kj} dy^j - \sum_{k=o}^{m} \sum_{j=m+1}^{n} y^k \bar{\omega}_{kj} dy^j$$

$$= \sum_{k=o}^{m} \bar{\omega}_k dy^k,$$

where $\bar{\omega}_{kj} = \partial'_j \bar{\omega}^k$. This implies that $\bar{\omega}_j = 0$ for $m+1 \leq j \leq n$. Thus, $\bar{\omega} = \sum_{a=o}^{m} \bar{\omega}_a dy^a$.

Notice that $\Gamma_{\omega'}(T_N)$ contains a suitable extension of $\partial'_{m+i} - y^i\partial'_o$. Since

$$\mathcal{L}_{\partial'_{m+i} - y^i\partial'_o} \bar{\omega} = \sum_{j=o}^{m} \partial'_{m+i}\bar{\omega}_j \, dy^j + \bar{\omega}_o \, dy^i = \bar{\tau}_i\bar{\omega},$$

we have

$$\partial'_{m+i}\bar{\omega}_o = \bar{\tau}_i\bar{\omega}_o, \ \ldots, \ \partial'_{m+i}\bar{\omega}_i + \bar{\omega}_o = \bar{\tau}_i\bar{\omega}_i, \ \ldots, \ \partial'_{m+i}\bar{\omega}_m = \bar{\tau}_i\bar{\omega}_m .$$

Solve these equations, and we see that $(\exp - \int \bar{\tau}_i \, dy^{m+i})\bar{\omega}$ does not contain the variable y^{m+i} except i-th term. Thus, we can find a local C^∞-function τ such that $\tau\bar{\omega} = Cdy^o + f_i(y^{m+i})dy^i$. Since $C \neq 0$, we may put $C = 1$. Again we consider the Lie derivative

$$\mathcal{L}_{\partial'_{m+i} - y^i \partial'_o}(dy^o + f_i(y^{m+i})dy^i).$$

This is equal to $-dy^i + \frac{d}{dy^{m+i}} f(y^{m+i})dy^i$, but this can not be the shape $\tau\omega'$. Therefore, this must be 0, hence $\frac{d}{dy^{m+i}} f(y^{m+i}) = 1$. So, we have that $\tau\omega = dy^o + \sum_{i=1}^{m} y^{m+i}dy^i$.

Now, what we get is that for every point $p \in N$, there is a C^∞-function τ_p defined on a neighborhood of p such that $\tau_p\omega' = dy^o + \sum_{i=1}^{m} y^{m+i}dy^i$. Therefore, we see that $\varphi^*(\tau_p\omega') = (\varphi^*\tau_p)\varphi^*\omega' = \omega$. Since ω and $\varphi^*\omega'$ are globally defined 1-form on M, we see that $\varphi^*\tau_p$ is also globally defined smooth function on M, and so is τ_p.

X.7 Proof of Theorem for the case $\alpha = \mathfrak{F}$.

Let B_M be the base space of the fibering and $\pi : M \to B_M$ the projection. Obviously, $\Gamma_{\mathfrak{F}}(T_M) = \{ u \in \Gamma(T_M) : d\pi u$ is constant on each fiber $\}$. (Therefore, we have to assume that the fibre is compact, because otherwise any $u \in \Gamma_{\mathfrak{F}}(T_M)$ with compact support must satisfy $d\pi u = 0$.) The natural mapping $d\pi : \Gamma_{\mathfrak{F}}(T_M) \to \Gamma(T_{B_M})$ is a surjective homomorphism as Lie algebras. Let $\Gamma_F(T_M)$ be its kernel.

Now, consider what is a standard coordinate of such fibering. This must be a local coordinate $x^1,\ldots,x^k,x^{k+1},\ldots,x^{k+\ell}$ such that for any fixed $x^1,\ldots,x^k$, a local coordinate of the fibre is given by $x^{k+1},\ldots,x^{\ell}$. However, the following is also easy to prove.

10.7.1 Lemma <u>If</u> $u \in \Gamma_{\mathfrak{F}}(T_M)$ <u>satisfies</u> $u(p) \neq 0$, <u>then there is a standard coordinate</u> $x^1,\ldots,x^k,x^{k+1},\ldots,x^{k+\ell}$ at p such that $u \equiv \partial_1$ on a neighborhood of p, or

$u = \partial_{k+1} + \sum_{i=1}^{k} f^i(x^1,\ldots,x^k)\,\partial_i$ with $f^i(0,\ldots,0) = 0$, where as a matter of course the origin of the coordinate corresponds to the point p.

Let $\mathcal{J}$ be an ideal of $\Gamma_{\mathcal{F}}(T_M)$ such that for any point $p \in M$, there us $u \in \mathcal{J}$ such that $(d\pi u)(\pi p) \neq 0$. Then, by 10.3.2 Lemma and Proposition 2, we see that $\mathcal{J} + \Gamma_F(T_M) = \Gamma_{\mathcal{F}}(T_M)$. However, since $\mathcal{J}$ contains a suitable extension of ∂_1 in a standard coordinate at every point $p \in M$, $\mathcal{J}$ must contain any element $v \in \Gamma_F(T_M)$ such that the support of v is contained in that coordinate neighborhood. Therefore, we see that $\mathcal{J} \supset \Gamma_F(T_M)$. Thus, $\mathcal{J} = \Gamma_{\mathcal{F}}(T_M)$.

By the above argument, we see that for any ideal $\mathcal{J} \subsetneq \Gamma_{\mathcal{F}}(T_M)$, there is a point $\tilde{p} \in B_M$ such that every element $d\pi u$ of $d\pi\mathcal{J}$ vanishes at $\tilde{p}$ with all of its derivatives. Hence we get

10.7.2 Lemma Every maximal ideal of $\Gamma_{\mathcal{F}}(T_M)$ contains the ideal $\Gamma_F(T_M)$. Moreover, this must be equal to $d\pi^{-1}\mathcal{J}_{\tilde{p}}$ for a point $\tilde{p} \in B_M$, where $\mathcal{J}_{\tilde{p}}$ is the maximal ideal of $\Gamma(TB_M)$ corresponding to the point $\tilde{p}$.

Now, we can characterize the ideal $\Gamma_F(T_M)$. This is given by the intersection of all maximal ideal of $\Gamma_{\mathcal{F}}(T_M)$.

10.7.3 Lemma Let $\mathcal{J}$ be an ideal of $\Gamma_{\mathcal{F}}(T_M)$ such that for any point $p \in M$, there is an element $u \in \mathcal{J}$ such that $u(p) \neq 0$. Then, $\mathcal{J}$ must contain the ideal $\Gamma_F(T_M)$.

Proof. For every point $p \in M$, we choose a standard coordinate $x^1,\ldots,x^k,x^{k+1},\ldots,x^{k+\ell}$, by using 10.7.1. It is enough to show that $\mathcal{J}$ contains every element $v \in \Gamma_F(T_M)$ such that the support of v is contained in the coordinate neighborhood.

If $(d\pi u)(\pi p) \neq 0$, then this case is easy and in fact has been already considered So, suppose $(d\pi u)(\pi p) = 0$. Then, we may assume that $u = \partial_{k+1} + \sum_{i=1}^{k} f^i(x^1,\ldots,x^k)\partial_i$. Notice that $\Gamma_{\mathcal{F}}(T_M)$ contains a suitable extension of $y^{k+1}\partial_{k+1}$. Hence $\partial_{k+1} = [u, x^{k+1}\partial_{k+1}]$ is also contained in $\mathcal{J}$. Thus, we can use the same argument as in § X,3 and obtain that $\mathcal{J}$ must contain any element $v \in \Gamma_F(T_M)$.

Now, consider an ideal $\mathcal{I}$ of $\Gamma_{\mathcal{F}}(T_M)$ such that $\mathcal{I} \subsetneqq \Gamma_F(T_M)$ and maximal under these conditions. We call such an ideal a maximal ideal of $\Gamma_{\mathcal{F}}(T_M)$ in $\Gamma_F(T_M)$. By the previous lemma, there is a point $p \in M$ such that every element of $\mathcal{I}$ vanishes at p together with all derivatives. Moreover, this point is uniquely determined. We denote by $\mathcal{J}_p$ the maximal ideal of $\Gamma_{\mathcal{F}}(T_M)$ in $\Gamma_F(T_M)$ corresponding to $p \in M$, and by $\tilde{\mathcal{I}}_{\tilde{p}}$ the maximal ideal of $\Gamma_{\mathcal{F}}(T_M)$ given by $d\pi^{-1}\mathcal{I}_{\tilde{p}}$, $\tilde{p} \in B_M$.

The above arguments show that an isomorphism $\Phi : \Gamma_{\mathcal{F}}(T_M) \rightarrow \Gamma_{\mathcal{F}}(T_N)$ yields a homeomorphism $\varphi : M \rightarrow N$. (Cf. § X.1.)

To prove φ is a $\mathcal{F}$-preserving diffeomorphism, we have to characterize the zeros of a vector field $u \in \Gamma_{\mathcal{F}}(T_M)$. (Cf.Lemma 3.)

10.7.4 Lemma <u>For an element</u> $u \in \Gamma_{\mathcal{F}}(T_M)$, $d\pi(u(p)) \neq 0$ <u>if and only if</u> $[u,\Gamma_{\mathcal{F}}(T_M)] + \tilde{\mathcal{I}}_{\pi p} = \Gamma_{\mathcal{F}}(T_M)$. <u>Moreover, under the assumption</u> $d\pi(u(p)) = 0$, $u(p) \neq 0$ <u>if and only if</u> $[u,\Gamma_F(T_M)] + \mathcal{J}_p = \Gamma_F(T_M)$.

Proof. The first one is trivial by using Lemma 3. For the second one, we may put $u = \partial_{k+1} + \sum_{i=1}^{k} f^i(x^1,\dots,x^k)\partial_i$. Since $\partial_{k+1}h + \sum_{i=1}^{k} f^i(x^1,\dots,x^k)\partial_i h = g$ can be solved locally for given g, we can use the same argument as in §X.5 and get the desired result.

Now, we can prove the theorem for the case $\alpha = \mathcal{F}$. Since for any point p, $\Gamma_{\mathcal{F}}(T_M)$ contains suitable extensions of $\partial_1,\dots,\partial_k, \partial_{k+1},\dots,\partial_{k+\ell}$, we put $u_i = \Phi(\partial_i)$, $v_i = \Phi(\partial_{k+i})$. Then easily $[u_i,u_j] = [v_i,v_j] = 0$ on a neighborhood of $\varphi(p)$. Thus, there is a coordinate $y^1,\dots,y^k,y^{k+1},\dots,y^{k+\ell}$ at $\varphi(p)$ such that $u_i = \partial'_i$ and $v_i = \partial'_{k+i}$.

Notice that $\Gamma_{\mathcal{F}}(T_M)$ contains also suitable extensions of $x^i\partial_i$, $x^{k+i}\partial_{k+i}$ and remark that $[\partial_i, x^i\partial_i] = \partial_i$ etc.. Then, by the same argument as in §X.6 shows that $y^i(\varphi(*)) = x^i(*)$, $y^{k+i}(\varphi(*)) = x^{k+i}(*)$. Since $v_1,\dots,v_\ell$ are contained in $\Gamma_{F'}(T_N)$, we see that φ is a fibre preserving diffeomorphism.

§ XI Linear groups and groups of diffeomorphisms

XI.1 Linear groups $GL_o(E)$, $GL_{oo}(E)$.

Let $\{\mathbb{E}, E^k, k \in N(d)\}$ be a Sobolev chain. We denote by $L(\mathbb{E})$ the totality of continuous linear mapping of $\mathbb{E}$ into itself and by $GL(\mathbb{E})$ the totality of invertible element in $L(\mathbb{E})$. $GL(\mathbb{E})$ is, however, too huge to treat. For instance it seems to be difficult to define a topological group structure for $GL(\mathbb{E})$ except the trivial one, and this was the main reason why Frechet manifolds are difficult to handle with.

So, we consider a smaller class. Let $L_o(\mathbb{E})$ be the totality of $A \in L(\mathbb{E})$ such that A can be extended to a bounded linear operator of E^k into itself for every $k \in N(d)$, and let $L_{oo}(\mathbb{E})$ the totality of $A \in L_o(\mathbb{E})$ such that A satisfies the inequality

$$\|Au\|_k \leqslant C\|u\|_k + D_k\|u\|_{k-1} \quad \text{for } k \geqslant d+1,$$

where C, D_k are positive constants and C does not depend on k.

$GL_o(\mathbb{E})$ (resp. $GL_{oo}(\mathbb{E})$) be the totality of elements $A \in L_o(\mathbb{E})$ (resp. $L_{oo}(\mathbb{E})$) such that A^{-1} exists and is contained in $L_o(\mathbb{E})$ (resp. $L_{oo}(\mathbb{E})$). The purpose of this section is to define several topologies for $GL_o(\mathbb{E})$ and $GL_{oo}(\mathbb{E})$.

(A) PLN-topology for $L_o(\mathbb{E})$.

Since $A \in L_o(\mathbb{E})$ can be extended to a bounded operator of E^k into itself for any $k \in N(d)$. We can define the norm $\|A\|_k$ of A in $L(E^k, E^k)$. Thus, we can consider on $L_o(\mathbb{E})$ the <u>P</u>rojective <u>L</u>imit topology of the <u>N</u>orm topology on $L(E^k, E^k)$.

$L_o(\mathbb{E})$ is a projective limit of Banach spaces.

$GL_o(\mathbb{E})$ with relative topology is then a topological group. We can easily define an ILB-Lie group structure on $GL_o(\mathbb{E})$. However, it seems to be impossible to prove that $GL_o(\mathbb{E})$ is an open subset of $L_o(\mathbb{E})$.

(B) Strong PLN-topology for $L_{oo}(E)$.

Since $A \in L_{oo}(E)$ satisfies the inequality $\|Au\|_k \leqslant C\|u\|_k + D_k\|u\|_{k-1}$, we denote by $|A|$ the infimum of such possible constant C. Then, it is easy to see the following :

11.1.1 Lemma $|A|$ <u>is a semi-norm satisfying</u> $|AB| \leq |A|\cdot|B|$.

The strong PLN-topology for $L_{oo}(\mathbb{E})$ is defined by semi-norm $|\ |$ and operator norms of $L(E^k,E^k)$ as a projective limit topology.

11.1.2 Theorem $GL_{oo}(\mathbb{E})$ <u>is an open subset of</u> $L_{oo}(\mathbb{E})$ <u>with the strong PLN-topology.</u>

Proof. Let $|||A|||_k$ be the maximum of $|A|$, $\|A\|_d$, ..., $\|A\|_k$, where $\|A\|_k$ is the operator norm in $L(E^k,E^k)$. Let $L^k_{oo}(\mathbb{E})$ be the completion of $L_{oo}(\mathbb{E})$ by the norm $|||\ |||_k$. Then, $L^k_{oo}(\mathbb{E})$ is a Banach algebra and $L_{oo}(\mathbb{E}) = \cap L^k_{oo}(\mathbb{E})$. Obviously, $L^k_{oo}(\mathbb{E}) \subset L^{k-1}_{oo}(\mathbb{E})$ and the inclusion is continuous.

We put $W = \{ I + B \in L^d_{oo}(\mathbb{E}) : |||B|||_d < 1, |B| < \frac{1}{2} \}$. We have only to show that $W \cap L_{oo}(\mathbb{E}) \subset GL_{oo}(\mathbb{E})$. We use the induction.

It is easy to see that $W \subset GL^d_{oo}(\mathbb{E})$, (i.e. the invertible elements in $L^d_{oo}(\mathbb{E})$.) Assume that $W \cap L^{k'}_{oo}(\mathbb{E}) \subset GL^{k'}_{oo}(\mathbb{E})$ for any $d \leq k' < k$. Since $L^k_{oo}(\mathbb{E})$ is a Banach algebra, we see that $W \cap GL^k_{oo}(\mathbb{E})$ is an open subset of $W \cap L^k_{oo}(\mathbb{E})$. So, to prove $W \cap GL^k_{oo}(\mathbb{E}) = W \cap L^k_{oo}(\mathbb{E})$, we have only to show that $W \cap GL^k_{oo}(\mathbb{E})$ is closed in $W \cap L^k_{oo}(\mathbb{E})$, because $W \cap L^k_{oo}(\mathbb{E})$ is a connected set.

Let A be a boundary point of $W \cap GL^k_{oo}(\mathbb{E})$ in $W \cap L^k_{oo}(\mathbb{E})$. We choose a sequence $\{A_n\}$ such that $A_n \in W \cap GL^k_{oo}(\mathbb{E})$ and $\lim A_n = A$.

11.1.3 Lemma $A : E^k \mapsto E^k$ <u>is an isomorphism.</u>

Proof. Since we know that $A : E^k \mapsto E^k$ is injective (because $A : E^d \mapsto E^d$ is an isomorphism,) we have only to show that $AE^k = E^k$. Since $A \in W \cap L^k_{oo}(\mathbb{E})$, we see

$$\|u\|_k - \|Au\|_k \leq \|(I - A)u\|_k \leq \frac{1}{2} \|u\|_k + D_k\|u\|_{k-1} ,$$

so that $\|Au\|_k \geq \frac{1}{2} \|u\|_k - D_k\|u\|_{k-1}$. Since $A \in GL^{k-1}_{oo}(\mathbb{E})$, this inequality shows that the image of A is closed in E^k.

Now, let y be an arbitrary point in E^k. Then, we have $x_n \in E^k$ such that $A_n x_n = y$. We have only to show that Ax_n converges to y in E^k.

$$\|A_n u\|_k \geq \|Au\|_k - \|A_n - A\|_k \|u\|_k \geq \frac{1}{2}\|u\|_k - D_k\|u\|_{k-1} - \|A_n - A\|_k\|u\|_k .$$

Thus, for sufficiently large n, we see

$$\|A_n u\|_k \geq \frac{1}{3}\|u\|_k - D_k\|u\|_k .$$

Put $u = x_n$, and use the assumption of the induction. Since $\|x_n\|_{k-1}$ is bounded, we see that $\{\|x_n\|_k\}$ is also bounded. Since $\|y - Ax_n\|_k = \|A_n x_n - Ax_n\|_k \leq \|A_n - A\|_k \|x_n\|_k$ we see that Ax_n converges to y.

By the above lemma, we have that $A : E^k \to E^k$ is an isomorphism. Moreover, by the inequality $\|Au\|_k \geq \frac{1}{2}\|u\|_k - D_k\|u\|_{k-1}$, we see that $\|u\|_k \geq \frac{1}{2}\|A^{-1}u\|_k - D_k\|A^{-1}u\|_{k-1}$ and hence $\|A^{-1}u\|_k \leq 2\|u\|_k + D_k'\|u\|_{k-1}$. This implies $A \in GL^k_{oo}(\mathbb{E})$. Thus, we get $W \cap GL^k_{oo}(\mathbb{E}) = W \cap L^k_{oo}(\mathbb{E})$ for every $k \in N(d)$. This completes the proof.

Remark $GL^k_{oo}(\mathbb{E})$ is a strong ILB-Lie group. Moreover, every $GL^k_{oo}(\mathbb{E})$ is a Banach Lie group.

(C) PLS-topology for $GL_o(\mathbb{E})$.

Here we use a slight modification of the strong topology. We call a sequence $\{A_n\}$ in $GL(E^k)$ converges to $A \in GL(E^k)$, if and only if s-lim$A_n = A$ and s-lim$A_n^{-1} = A^{-1}$, where s-lim means the limit in the strong topology of $L(E^k, E^k)$. This defines a topology on $GL(E^k)$, and by virtue of the uniformly boundedness theorem, we see that $GL(E^k)$ is a topological group. PLS-topology for $GL_o(\mathbb{E})$ is the projective limit topology of the above topologies for $GL(E^k)$.

The group $GL_o(E)$ with the PLS-topology has no manifold structure like a Frechet Lie group or a strong ILB-Lie group. However, this group containe many important groups.

Let $\{\Gamma(1_M), \Gamma^k(1_M), k \in N(d)\}$ be the Sobolev chain of functions defined in §II.2 where M is a closed, smooth manifold and $d = \dim M + 5$.

11.1.4 Theorem The group of diffeomorphism $\mathcal{D}(M)$ is a closed subgroup of $GL_o(\Gamma(1_M))$

with PLS-topology.

Proof. Let $r : \Gamma(1_M) \times \mathfrak{D} \mapsto \Gamma(1_M)$ be the mapping defined by $r(f,\varphi)(x) = f(\varphi(x))$. Then, by 2.1.2, we know that r can be extended to a C^{ℓ}-mapping of $\Gamma^{k+\ell}(1_M) \times \mathfrak{D}^k$ into $\Gamma^k(1_M)$. Especially r is a C^0-mapping of $\Gamma^k(1_M) \times \mathfrak{D}$ onto $\Gamma^k(1_M)$ for every $k \in N(d)$. Let $\rho(\varphi)f = r(f,\varphi^{-1})$ and we see that ρ is a mapping of $\mathfrak{D}$ into $GL_o(\Gamma(1_M))$. Moreover, the continuity of $r : \Gamma^k(1_M) \times \mathfrak{D} \mapsto \Gamma^k(1_M)$ implies that ρ is a continuous mapping of $\mathfrak{D}$ into $GL_o(\Gamma(1_M))$ with PLS-topology. Clearly, this is injective. Thus, we have a continuous monomorphism ρ of $\mathfrak{D}$ into $GL_o(\Gamma(1_M))$.

Moreover, each $\rho(\varphi)$ is a ring automorphism of $\Gamma(1_M)$. Therefore, if A is in the closure of $\rho(\mathfrak{D})$ in $GL_o(\Gamma(1_M))$, then A is also a ring automorphism. However, it is known that any ring automorphism of $\Gamma(1_M)$ is a C^∞-diffeomorphism. Namely, $(Af)(x) = f(\varphi(x))$ for a C^∞-diffeomorphism φ. Thus, we have the closedness of $\rho(\mathfrak{D})$.

XI.2 Linear groups which contains all groups of diffeomorphisms.

Let $\{\mathbb{E},E^k,k \in N(d)\},\{\mathbb{F},F^k,k \in N(d)\}$ be Sobolev chains. We call these Sobolev chains are equivalent (notation $\{\mathbb{E},E^k,k \in N(d)\}\sim\{\mathbb{F},F^k,k \in N(d)\}$), if there is a linear isomorphism ι of $\mathbb{E}$ onto $\mathbb{F}$ such that ι can be extended to an isomorphism of E^k onto F^k for every $k \in N(d)$.

$\{\mathbb{E},E^k,k \in N(d)\}$ and $\{\mathbb{F},F^k,k \in N(d)\}$ are called <u>normally equivalent</u> (notation $\simeq$) if they are equivalent and moreover ι satisfies the inequality

$$\|\iota u\|_k \leq C\|u\|_k + D_k\|u\|_{k-1}, \quad \|\iota^{-1}u\|_k \leq C'\|u\|_k + D_k'\|u\|_{k-1} \quad \text{for } k \in N(d+1),$$

where C, C', D_k, D_k' are positive constants and C, C' do not depend on k.

<u>Remark</u> In this article, if two Sobolev chain are normally equivalent, then they can be identified. However, even if two Sobolev chains are equivalent, there are many theorems in which we can not identify these. For instance groups $GL_{oo}(\mathbb{E})$ and $GL_{oo}(\mathbb{F})$ might be different. However, $GL_o(\mathbb{E})$ depends only on a equivalence class.

Let $\{\Gamma(1_M),\Gamma^k(1_M),k \in N(0)\}$ be the Sobolev chain of functions. We put $m = \dim M$ and define an abstract Sobolev chain $\{\mathbb{E},E^k,k \in N(0)\}$ as follows :

Put $E^k = \{(a_1, a_2, \dots) : \sum_{i=1}^{\infty} a_i^2 \, i^{\frac{2k}{m}} < \infty\}$. Then, we see that E^k is a Hilbert space with the norm $\|(a_1, a_2, \dots)\|_k^2 = \sum_{i=1}^{\infty} a_i^2 \, i^{\frac{2k}{m}}$. We put $\mathbb{E} = \cap E^k$ with the inverse limit topology. Obviously, $\{\mathbb{E}, E^k, k \in N(0)\}$ is a Sobolev chain.

11.2.1 Theorem $\{\Gamma(1_M), \Gamma^k(1_M), k \in N(0)\}$ <u>is equivalent to</u> $\{\mathbb{E}, E^k, k \in N(0)\}$.

Proof. Define an arbitrary smooth riemannian metric on M. Let Δ be the Laplace operator with respect to the riemannian metric. We can consider $P = (\Delta+1)^{\frac{1}{2}}$ in the sense of singular integral operators. This is known to be an isomorphism of $\Gamma^k(1_M)$ onto $\Gamma^{k-1}(1_M)$ for any $k \geq 1$.

Let $e_1, e_2, \dots, e_n, \dots$ be the eigen functions of $\Delta + 1$. Then these are also eigen functions of P. We normalize these by the condition $\langle e_i, e_j \rangle_0 = \delta_{ij}$.

Let $0 < \lambda_1 \leq \lambda_2 \leq \cdots$ be the series of eigen values of P, where we take multiplicities into account. We may assume that every e_i is an eigen function corresponding to the eigen value λ_i.

Now, it is well-known that $\lim_{i \to \infty} \frac{\lambda_i^m}{i}$ exists and is positive. (This number is related to the total volume of M.) Define a mapping $\iota : \Gamma^0(1_M) \to E^0$ by $\iota(e_i) = \varepsilon_i$ where $\varepsilon_i = (\underbrace{0, \dots, 0, 1}_{i}, 0, \dots)$. Then, ι is an linear isometry.

We have only to show that the restriction of ι onto $\Gamma^k(1_M)$ is an isomorphism of $\Gamma^k(1_M)$ onto E^k. Remark that any element $u \in \Gamma^k(1_M)$ is expressed by $u = \sum a_i e_i$, $\sum a_i^2 \lambda_i^{2k} < \infty$. Thus, $\iota(u) = \sum a_i \varepsilon_i$. To prove this is contained in E^k is to show that $\sum a_i^2 \, i^{\frac{2k}{m}} < \infty$. Since $\lim (\lambda_i^m / i)$ exists and positive, we see that there are positive constants K_k, K'_k such that

$$K_k \sum a_i^2 \, i^{\frac{2k}{m}} \leq \sum a_i^2 \lambda_i^{2k} \leq K'_k \sum a_i^2 \, i^{\frac{2k}{m}} .$$

This implies $\iota u \in E^k$ and $\iota : \Gamma^k(1_M) \mapsto E^k$ is an isomorphism.

11.2.2 Corollary <u>Let</u> N <u>be another smooth, closed manifold such that</u> $\dim N \leq \dim M = m$. <u>Then,</u> $\mathcal{D}(N)$ <u>is embedded in</u> $GL_0(\Gamma(1_M))$ <u>as a closed subgroup.</u>

Proof. Let $n = \dim N$. Then, the Sobolev chain $\{\Gamma(1_N), \Gamma^k(1_N), k \in N(d)\}$, $d = \dim M + 5$, is equivalent to $\{\mathbb{F}, F^k, k \in N(d)\}$ where $F^k = \{(a_1, a_2, \dots) : \sum a_i^2 i^{\frac{2k}{n}} < \infty\}$ with the natural Hilbert norm and $\mathbb{F} = \cap F^k$.

Since $n \leq m$, there is a subsequence $s_1, s_2, \dots$ of the integers such that

$$0 < \underline{\lim}\ i^{\frac{1}{n}}/s_i^{\frac{1}{m}} \leq \overline{\lim}\ i^{\frac{1}{n}}/s_i^{\frac{1}{m}} < \infty .$$

Let E_1^k be the closure of the subspace of E^k spaned by $\varepsilon_{s_1}, \varepsilon_{s_2}, \dots$. We put $\mathbb{E}_1 = \cap E_1^k$. Then, obviously, $\{\mathbb{E}_1, E_1^k, k \in N(d)\}$ is equivalent to $\{\mathbb{F}, F^k, k \in N(d)\}$. Let E_2^k be the closure of the subspace of E^k spaned by the complement of the set $\{\varepsilon_{s_1}, \varepsilon_{s_2}, \dots\}$. Then, E_1^k is perpendicular to E_2^k. Hence $E^k = E_1^k \oplus E_2^k$ and $\mathbb{E} = \mathbb{E}_1 \oplus \mathbb{E}_2$. Remark that $\mathfrak{D}(N)$ is a closed subgroup of $GL_o(\mathbb{E}_1)$ with PLS-topology. Since $GL_o(\mathbb{E}_1)$ is natually imbedded in $GL_o(\mathbb{E})$ as a closed subgroup, we have the desired result.

Now, let Γ^k be the Hilbert space defined by

$$\{ (a_1, a_2, \dots) : \sum a_i^2 (\log i)^{2k} < \infty \}$$

with the natural Hilbert norm. Put $\mathbb{\Gamma} = \cap \Gamma^k$ with the inverse limit topology. Then, $\{\mathbb{\Gamma}, \Gamma^k, k \in N(d)\}$ is a Sobolev chain for any $d \geq 0$. Since we have to consider $GL_o(\mathbb{\Gamma})$ of $\{\mathbb{\Gamma}, \Gamma^k, k \in N(d)\}$ for different d, we indicate this by attaching a suffix d ; $GL_o(\mathbb{\Gamma}, d)$.

Obviously,

$$GL_o(\mathbb{\Gamma}, 0) \subset GL_o(\mathbb{\Gamma}, 1) \subset GL_o(\mathbb{\Gamma}, 2) \subset \cdots .$$

Let $GL_o(\mathbb{\Gamma}, \infty) = \cup GL_o(\mathbb{\Gamma}, d)$. We consider the PLS-topology for $GL_o(\mathbb{\Gamma}, d)$ and the inductive limit topology for $GL_o(\mathbb{\Gamma}, \infty)$ (briefly, ILPLS-topology.) Namely, a sequence $\{A_n\} \in GL_o(\mathbb{\Gamma}, \infty)$ is said to converge to $A \in GL_o(\mathbb{\Gamma}, \infty)$ if and only if there is d such that A and A_n for sufficiently large n are contained in $GL_o(\mathbb{\Gamma}, d)$ and $\{A_n\}$ converges to A in $GL_o(\mathbb{\Gamma}, d)$. $GL_o(\mathbb{\Gamma}, \infty)$ with the above topology is a topological group.

The following is now easy to prove by the same method as in 11.2.2.

11.2.3 Theorem $GL_0(\mathbb{C},\infty)$ with ILPLS-topology contains groups of diffeomorphisms of all closed manifolds as closed subgroups.

This paper was written under the program of the "Sonderforschungsberei Theoretische Mathematik" of Bonn University.

University of Bonn and Tokyo Metropolitan University.

Present adress:

Dept. Math.
Tokyo Metropolitan University,
Fukazawa, Setagaya,
Tokyo, Japan.

References

[1] R.Abraham, Foundation of mechanics, Benjamin N.Y. 1967.

[2] R.Abraham and J.Robbin, Transversal mappings and flows, Benjamin N.Y. 1967.

[3] W.M.Boothby and H.Wang, On contact manifolds, Ann.Math. 68 (1958) 721-734.

[4] N.Bourbaki, Groupes et algebres de Lie, Chap I. Hermann 1960.

[5] E.Calabi, On differential action of compact Lie group on compact manifolds, Proc. conference on transformation groups, Springer 1968 210-213.

[6] S.S.Chern, The geometry of G-structures, Bull.Amer.Math.Soc.72,1966,167-219.

[7] P.de la Harp, Classical Banach-Lie algebras and Banach Lie groups, Lecture note, Springer,285, 1972.

[8] J.Dieudonne, Foundations of modern analysis, Acad.Press, 1960.

[9] J.Eells Jr., A setting for global analysis, Bull.Amer.Math.Soc.72,1966,751-807.

[10] D.Ebin and J.Marsden, Groups of diffeomorphisms and the motion of an imcompressible fluid, Ann.Math. 1970, 101-162.

[11] D.Ebin, Espace des metriques Riemanniens et mouvement des fluides via les varietes d'applications, Lecture note l'Ecole Polytechnique et Univ. Paris, VII,1972.

[12] D.Ebin, J.Marsden and A.Fisher, Diffeomorphism groups, hydrodynamics and relativity, Proc.congress on diff. topology and diff. geometry, Cnad.Math.Soc. 1972, 135-279.

[13] J.W.Gray, Some global properties of contact structures, Ann.Math.69(1959)421-450.

[14] A.M.Gleason and R.S.Palais, On a class of transformation groups, Amer.J.Math.79 (1957) 631-648.

[15] V.W.Guillemin, Infinite dimensional primitive Lie algebras, J.Diff.Geom.(1970) 257-

[16] L.Hörmander, On interior regularity of the solution of partial differential equations, Comm.Pure Appl.Math. XI (1958) 197-218.

[17] S.Kobayashi and T.Nagano, On filtered Lie algebras and geometric structures, III, IV, J.Math.Mech. 14(1956) 679-706 and 15(1966) 163-175.

[18] N.H.Kuiper, The homotopy type of the unitary group of Hilbert space, Topology 3

(1965) 19-30.

[19] M.Kuranishi, On the local theory of continuous infinite pseudo groups, I, II, Nagoya Math.J. 15 (1959) 225-260 and 19(1961) 55-91.

[20] J.Leslie, On a differentiable structures of groups of diffeomorphisms, Topology 6 (1967).

[21] -------, Some Frobenius theorems in global analysis, J.Diff.Geom.1968,279-298.

[22] S.Mizohata, 偏微分方程式論, Iwanami (岩波) 1965.

[23] T.Morimoto and N.Tanaka, The classification of real primitive infinite Lie algebras, J.Math.Kyoto Univ. 10, 1970, 207-243.

[24] D.Montgomery and L.Zippin, Topological transformation groups, Interscience N.Y. 1955.

[25] J.Moser, A new technique for construction of solutions of non-linear differential equations, Proc.Nat.Acad.Sci.U.S.A. 47 (1961).

[26] E.Nelson, Topics in dynamics I. flows, Math. Note, Princeton Press 1969.

[27] H.Omori, Homomorphic images of Lie groups, J.Math.Soc.Japan,18,1966,97-117.

[28] -------, On the group of diffeomorphisms on a compact manifold, Proc.Symp.Pure Appl.Math. XV. Amer.Math.Soc.1970, 167-183.

[29] -------, On regularity of connections, Differential Geometry, dedicated to K.Yano, Kinokuniya, 1972.

[30] ------, Local structures of groups of diffeomorphisms, J.Math.Soc.Japan 24,1972, 60-88.

[31] ------, On smooth extension theorems,J.Math.Soc.Japan, 24, 1972, 405-432.

[32] ------, Groups of diffeomorphisms and their subgroups, Trans.Amer.Math.Soc. 1973, to appear.

[33] ------ and P. de la Harpe, About interaction between Banach Lie groups and finite dimensional manifolds, J.Math.Kyoto Univ. 12-3, 1972.

[34] ------, On Lie algebras of vector fields, to appear.

[35] R.S.Palais, Equivalence of nearby differential actions of compact Lie groups, Bull Amer.Math.Soc. 67 (1961) 362-364.

[36] --------, et al., Seminor on Atiyah - Singer index theorem, Princeton Study 57,

1965.

[37] R.S.Palais, Homotopy theory of infinite dimensional manifolds, Topology 5,1966, 1-16.

[38] J.Peetre, Une caractérisation abstraite des opérateurs differentiels, Math.Scand. 7,1959, 211-218.

[39] L.E.Pursell and Shanks, The Lie algebra of smooth manifold, Proc.Amer.Math.Soc. 1954, 468-472.

[40] S.Shnider, The classification of real primitive infinite Lie algebras, J.Diff. Geom. 4 (1970) 81-89.

[41] I.M.Singer and S.Sternberg, On the infinite groups of Lie and Cartan I, J.Anal. Math. 15, 1965, 1-114.

Vol. 247: Lectures on Operator Algebras. Tulane University Ring and Operator Theory Year, 1970–1971. Volume II. XI, 786 pages. 1972. DM 40,–

Vol. 248: Lectures on the Applications of Sheaves to Ring Theory. Tulane University Ring and Operator Theory Year, 1970–1971. Volume III. VIII, 315 pages. 1971. DM 26,–

Vol. 249: Symposium on Algebraic Topology. Edited by P. J. Hilton. VII, 111 pages. 1971. DM 16,–

Vol. 250: B. Jónsson, Topics in Universal Algebra. VI, 220 pages. 1972. DM 20,–

Vol. 251: The Theory of Arithmetic Functions. Edited by A. A. Gioia and D. L. Goldsmith VI, 287 pages. 1972. DM 24,–

Vol. 252: D. A. Stone, Stratified Polyhedra. IX, 193 pages. 1972. DM 18,–

Vol. 253: V. Komkov, Optimal Control Theory for the Damping of Vibrations of Simple Elastic Systems. V, 240 pages. 1972. DM 20,–

Vol. 254: C. U. Jensen, Les Foncteurs Dérivés de $\varprojlim$ et leurs Applications en Théorie des Modules. V, 103 pages. 1972. DM 16,–

Vol. 255: Conference in Mathematical Logic – London '70. Edited by W. Hodges. VIII, 351 pages. 1972. DM 26,–

Vol. 256: C. A. Berenstein and M. A. Dostal, Analytically Uniform Spaces and their Applications to Convolution Equations. VII, 130 pages. 1972. DM 16,–

Vol. 257: R. B. Holmes, A Course on Optimization and Best Approximation. VIII, 233 pages. 1972. DM 20,–

Vol. 258: Séminaire de Probabilités VI. Edited by P. A. Meyer. VI, 253 pages. 1972. DM 22,–

Vol. 259: N. Moulis, Structures de Fredholm sur les Variétés Hilbertiennes. V, 123 pages. 1972. DM 16,–

Vol. 260: R. Godement and H. Jacquet, Zeta Functions of Simple Algebras. IX, 188 pages. 1972. DM 18,–

Vol. 261: A. Guichardet, Symmetric Hilbert Spaces and Related Topics. V, 197 pages. 1972. DM 18,–

Vol. 262: H. G. Zimmer, Computational Problems, Methods, and Results in Algebraic Number Theory. V, 103 pages. 1972. DM 16,–

Vol. 263: T. Parthasarathy, Selection Theorems and their Applications. VII, 101 pages. 1972. DM 16,–

Vol. 264: W. Messing, The Crystals Associated to Barsotti-Tate Groups: With Applications to Abelian Schemes. III, 190 pages. 1972. DM 18,–

Vol. 265: N. Saavedra Rivano, Catégories Tannakiennes. II, 418 pages. 1972. DM 26,–

Vol. 266: Conference on Harmonic Analysis. Edited by D. Gulick and R. L. Lipsman. VI, 323 pages. 1972. DM 24,–

Vol. 267: Numerische Lösung nichtlinearer partieller Differential- und Integro-Differentialgleichungen. Herausgegeben von R. Ansorge und W. Törnig, VI, 339 Seiten. 1972. DM 26,–

Vol. 268: C. G. Simader, On Dirichlet's Boundary Value Problem. IV, 238 pages. 1972. DM 20,–

Vol. 269: Théorie des Topos et Cohomologie Etale des Schémas. (SGA 4). Dirigé par M. Artin, A. Grothendieck et J. L. Verdier. XIX, 525 pages. 1972. DM 50,–

Vol. 270: Théorie des Topos et Cohomologie Etale des Schémas. Tome 2. (SGA 4). Dirigé par M. Artin, A. Grothendieck et J. L. Verdier. V, 418 pages. 1972. DM 50,–

Vol. 271: J. P. May, The Geometry of Iterated Loop Spaces. IX, 175 pages. 1972. DM 18,–

Vol. 272: K. R. Parthasarathy and K. Schmidt, Positive Definite Kernels, Continuous Tensor Products, and Central Limit Theorems of Probability Theory. VI, 107 pages. 1972. DM 16,–

Vol. 273: U. Seip, Kompakt erzeugte Vektorräume und Analysis. IX, 119 Seiten. 1972. DM 16,–

Vol. 274: Toposes, Algebraic Geometry and Logic. Edited by. F. W. Lawvere. VI, 189 pages. 1972. DM 18,–

Vol. 275: Séminaire Pierre Lelong (Analyse) Année 1970–1971. VI, 181 pages. 1972. DM 18,–

Vol. 276: A. Borel, Représentations de Groupes Localement Compacts. V, 98 pages. 1972. DM 16,–

Vol. 277: Séminaire Banach. Edité par C. Houzel. VII, 229 pages. 1972. DM 20,–

Vol. 278: H. Jacquet, Automorphic Forms on GL(2). Part II. XIII, 142 pages. 1972. DM 16,–

Vol. 279: R. Bott, S. Gitler and I. M. James, Lectures on Algebraic and Differential Topology. V, 174 pages. 1972. DM 18,–

Vol. 280: Conference on the Theory of Ordinary and Partial Differential Equations. Edited by W. N. Everitt and B. D. Sleeman. XV, 367 pages. 1972. DM 26,–

Vol. 281: Coherence in Categories. Edited by S. Mac Lane. VII, 235 pages. 1972. DM 20,–

Vol. 282: W. Klingenberg und P. Flaschel, Riemannsche Hilbertmannigfaltigkeiten. Periodische Geodätische. VII, 211 Seiten. 1972. DM 20,–

Vol. 283: L. Illusie, Complexe Cotangent et Déformations II. VII, 304 pages. 1972. DM 24,–

Vol. 284: P. A. Meyer, Martingales and Stochastic Integrals I. VI, 89 pages. 1972. DM 16,–

Vol. 285: P. de la Harpe, Classical Banach-Lie Algebras and Banach-Lie Groups of Operators in Hilbert Space. III, 160 pages. 1972. DM 16,–

Vol. 286: S. Murakami, On Automorphisms of Siegel Domains. V, 95 pages. 1972. DM 16,–

Vol. 287: Hyperfunctions and Pseudo-Differential Equations. Edited by H. Komatsu. VII, 529 pages. 1973. DM 36,–

Vol. 288: Groupes de Monodromie en Géométrie Algébrique. (SGA 7 I). Dirigé par A. Grothendieck. IX, 523 pages. 1972. DM 50,–

Vol. 289: B. Fuglede, Finely Harmonic Functions. III, 188. 1972. DM 18,–

Vol. 290: D. B. Zagier, Equivariant Pontrjagin Classes and Applications to Orbit Spaces. IX, 130 pages. 1972. DM 16,–

Vol. 291: P. Orlik, Seifert Manifolds. VIII, 155 pages. 1972. DM 16,–

Vol. 292: W. D. Wallis, A. P. Street and J. S. Wallis, Combinatorics: Room Squares, Sum-Free Sets, Hadamard Matrices. V, 508 pages. 1972. DM 50,–

Vol. 293: R. A. DeVore, The Approximation of Continuous Functions by Positive Linear Operators. VIII, 289 pages. 1972. DM 24,–

Vol. 294: Stability of Stochastic Dynamical Systems. Edited by R. F. Curtain. IX, 332 pages. 1972. DM 26,–

Vol. 295: C. Dellacherie, Ensembles Analytiques, Capacités, Mesures de Hausdorff. XII, 123 pages. 1972. DM 16,–

Vol. 296: Probability and Information Theory II. Edited by M. Behara, K. Krickeberg and J. Wolfowitz. V, 223 pages. 1973. DM 20,–

Vol. 297: J. Garnett, Analytic Capacity and Measure. IV, 138 pages. 1972. DM 16,–

Vol. 298: Proceedings of the Second Conference on Compact Transformation Groups. Part 1. XIII, 453 pages. 1972. DM 32,–

Vol. 299: Proceedings of the Second Conference on Compact Transformation Groups. Part 2. XIV, 327 pages. 1972. DM 26,–

Vol. 300: P. Eymard, Moyennes Invariantes et Représentations Unitaires. II. 113 pages. 1972. DM 16,–

Vol. 301: F. Pittnauer, Vorlesungen über asymptotische Reihen. VI, 186 Seiten. 1972. DM 18,–

Vol. 302: M. Demazure, Lectures on p-Divisible Groups. V, 98 pages. 1972. DM 16,–

Vol. 303: Graph Theory and Applications. Edited by Y. Alavi, D. R. Lick and A. T. White. IX, 329 pages. 1972. DM 26,–

Vol. 304: A. K. Bousfield and D. M. Kan, Homotopy Limits, Completions and Localizations. V, 348 pages. 1972. DM 26,–

Vol. 305: Théorie des Topos et Cohomologie Etale des Schémas. Tome 3. (SGA 4). Dirigé par M. Artin, A. Grothendieck et J. L. Verdier. VI, 640 pages. 1973. DM 50,–

Vol. 306: H. Luckhardt, Extensional Gödel Functional Interpretation. VI, 161 pages. 1973. DM 18,–

Vol. 307: J. L. Bretagnolle, S. D. Chatterji et P.-A. Meyer, Ecole d'été de Probabilités: Processus Stochastiques. VI, 198 pages. 1973. DM 20,–

Vol. 308: D. Knutson, λ-Rings and the Representation Theory of the Symmetric Group. IV, 203 pages. 1973. DM 20,–

Vol. 309: D. H. Sattinger, Topics in Stability and Bifurcation Theory. VI, 190 pages. 1973. DM 18,–

Vol. 310: B. Iversen, Generic Local Structure of the Morphisms in Commutative Algebra. IV, 108 pages. 1973. DM 16,–

Vol. 311: Conference on Commutative Algebra. Edited by J. W. Brewer and E. A. Rutter. VII, 251 pages. 1973. DM 22,–

Vol. 312: Symposium on Ordinary Differential Equations. Edited by W. A. Harris, Jr. and Y. Sibuya. VIII, 204 pages. 1973. DM 22,–

Vol. 313: K. Jörgens and J. Weidmann, Spectral Properties of Hamiltonian Operators. III, 140 pages. 1973. DM 16,–

Vol. 314: M. Deuring, Lectures on the Theory of Algebraic Functions of One Variable. VI, 151 pages. 1973. DM 16,–

Vol. 315: K. Bichteler, Integration Theory (with Special Attention to Vector Measures). VI, 357 pages. 1973. DM 26,–

Vol. 316: Symposium on Non-Well-Posed Problems and Logarithmic Convexity. Edited by R. J. Knops. V, 176 pages. 1973. DM 18,–

Vol. 317: Séminaire Bourbaki – vol. 1971/72. Exposés 400–417. IV, 361 pages. 1973. DM 26,–

Vol. 318: Recent Advances in Topological Dynamics. Edited by A. Beck, VIII, 285 pages. 1973. DM 24,–

Vol. 319: Conference on Group Theory. Edited by R. W. Gatterdam and K. W. Weston. V, 188 pages. 1973. DM 18,–

Vol. 320: Modular Functions of One Variable I. Edited by W. Kuyk. V, 195 pages. 1973. DM 18,–

Vol. 321: Séminaire de Probabilités VII. Edité par P. A. Meyer. VI, 322 pages. 1973. DM 26,–

Vol. 322: Nonlinear Problems in the Physical Sciences and Biology. Edited by I. Stakgold, D. D. Joseph and D. H. Sattinger. VIII, 357 pages. 1973. DM 26,–

Vol. 323: J. L. Lions, Perturbations Singulières dans les Problèmes aux Limites et en Contrôle Optimal. XII, 645 pages. 1973. DM 42,–

Vol. 324: K. Kreith, Oscillation Theory. VI, 109 pages. 1973. DM 16,–

Vol. 325: Ch.-Ch. Chou, La Transformation de Fourier Complexe et L'Equation de Convolution. IX, 137 pages. 1973. DM 16,–

Vol. 326: A. Robert, Elliptic Curves. VIII, 264 pages. 1973. DM 22,–

Vol. 327: E. Matlis, 1-Dimensional Cohen-Macaulay Rings. XII, 157 pages. 1973. DM 18,–

Vol. 328: J. R. Büchi and D. Siefkes, The Monadic Second Order Theory of All Countable Ordinals. VI, 217 pages. 1973. DM 20,–

Vol. 329: W. Trebels, Multipliers for (C, α)-Bounded Fourier Expansions in Banach Spaces and Approximation Theory. VII, 103 pages. 1973. DM 16,–

Vol. 330: Proceedings of the Second Japan-USSR Symposium on Probability Theory. Edited by G. Maruyama and Yu. V. Prokhorov. VI, 550 pages. 1973. DM 36,–

Vol. 331: Summer School on Topological Vector Spaces. Edited by L. Waelbroeck. VI, 226 pages. 1973. DM 20,–

Vol. 332: Séminaire Pierre Lelong (Analyse) Année 1971-1972. V, 131 pages. 1973. DM 16,–

Vol. 333: Numerische, insbesondere approximationstheoretische Behandlung von Funktionalgleichungen. Herausgegeben von R. Ansorge und W. Törnig. VI, 296 Seiten. 1973. DM 24,–

Vol. 334: F. Schweiger, The Metrical Theory of Jacobi-Perron Algorithm. V, 111 pages. 1973. DM 16,–

Vol. 335: H. Huck, R. Roitzsch, U. Simon, W. Vortisch, R. Walden, B. Wegner und W. Wendland, Beweismethoden der Differentialgeometrie im Großen. IX, 159 Seiten. 1973. DM 18,–

Vol. 336: L'Analyse Harmonique dans le Domaine Complexe. Edité par E. J. Akutowicz. VIII, 169 pages. 1973. DM 18,–

Vol. 337: Cambridge Summer School in Mathematical Logic. Edited by A. R. D. Mathias and H. Rogers. IX, 660 pages. 1973. DM 42,–

Vol: 338: J. Lindenstrauss and L. Tzafriri, Classical Banach Spaces. IX, 243 pages. 1973. DM 22,–

Vol. 339: G. Kempf, F. Knudsen, D. Mumford and B. Saint-Donat, Toroidal Embeddings I. VIII, 209 pages. 1973. DM 20,–

Vol. 340: Groupes de Monodromie en Géométrie Algébrique. (SGA 7 II). Par P. Deligne et N. Katz. X, 438 pages. 1973. DM 40,–

Vol. 341: Algebraic K-Theory I, Higher K-Theories. Edited by H. Bass. XV, 335 pages. 1973. DM 26,–

Vol. 342: Algebraic K-Theory II, "Classical" Algebraic K-Theory, and Connections with Arithmetic. Edited by H. Bass. XV, 527 pages. 1973. DM 36,–

Vol. 343: Algebraic K-Theory III, Hermitian K-Theory and Geometric Applications. Edited by H. Bass. XV, 572 pages. 1973. DM 38,–

Vol. 344: A. S. Troelstra (Editor), Metamathematical Investigation of Intuitionistic Arithmetic and Analysis. XVII, 485 pages. 1973. DM 34,–

Vol. 345: Proceedings of a Conference on Operator Theory. Edited by P. A. Fillmore. VI, 228 pages. 1973. DM 20,–

Vol. 346: Fučík et al., Spectral Analysis of Nonlinear Operators. II, 287 pages. 1973. DM 26,–

Vol. 347: J. M. Boardman and R. M. Vogt, Homotopy Invariant Algebraic Structures on Topological Spaces. X, 257 pages. 1973. DM 22,–

Vol. 348: A. M. Mathai and R. K. Saxena, Generalized Hypergeometric Functions with Applications in Statistics and Physical Sciences. VII, 314 pages. 1973. DM 26,–

Vol. 349: Modular Functions of One Variable II. Edited by W. Kuyk and P. Deligne. V, 598 pages. 1973. DM 38,–

Vol. 350: Modular Functions of One Variable III. Edited by W. Kuyk and J.-P. Serre. V, 350 pages. 1973. DM 26,–

Vol. 351: H. Tachikawa, Quasi-Frobenius Rings and Generalizations. XI, 172 pages. 1973. DM 18,–

Vol. 352: J. D. Fay, Theta Functions on Riemann Surfaces. V, 137 pages. 1973. DM 16,–

Vol. 353: Proceedings of the Conference on Orders, Group Rings and Related Topics. Organized by J. S. Hsia, M. L. Madan and T. G. Ralley. X, 224 pages. 1973. DM 20,–

Vol. 354: K. J. Devlin, Aspects of Constructibility. XII, 240 pages. 1973. DM 22,–

Vol. 355: M. Sion, A Theory of Semigroup Valued Measures. V, 140 pages. 1973. DM 16,–

Vol. 356: W. L. J. van der Kallen, Infinitesimally Central-Extensions of Chevalley Groups. VII, 147 pages. 1973. DM 16,–

Vol. 357: W. Borho, P. Gabriel und R. Rentschler, Primideale in Einhüllenden auflösbarer Lie-Algebren. V, 182 Seiten. 1973. DM 18,–

Vol. 358: F. L. Williams, Tensor Products of Principal Series Representations. VI, 132 pages. 1973. DM 16,–

Vol. 359: U. Stammbach, Homology in Group Theory. VIII, 183 pages. 1973. DM 18,–

Vol. 360: W. J. Padgett and R. L. Taylor, Laws of Large Numbers for Normed Linear Spaces and Certain Fréchet Spaces. VI, 111 pages. 1973. DM 16,–

Vol. 361: J. W. Schutz, Foundations of Special Relativity: Kinematic Axioms for Minkowski Space Time. XX, 314 pages. 1973. DM 26,–

Vol. 362: Proceedings of the Conference on Numerical Solution of Ordinary Differential Equations. Edited by D. Bettis. VIII, 490 pages. 1974. DM 34,–

Vol. 363: Conference on the Numerical Solution of Differential Equations. Edited by G. A. Watson. IX, 221 pages. 1974. DM 20,–

Vol. 364: Proceedings on Infinite Dimensional Holomorphy. Edited by T. L. Hayden and T. J. Suffridge. VII, 212 pages. 1974. DM 20,–

Vol. 365: R. P. Gilbert, Constructive Methods for Elliptic Equations. VII, 397 pages. 1974. DM 26,–

Vol. 366: R. Steinberg, Conjugacy Classes in Algebraic Groups (Notes by V. V. Deodhar). VI, 159 pages. 1974. DM 18,–

Vol. 367: K. Langmann und W. Lütkebohmert, Cousinverteilungen und Fortsetzungssätze. VI, 151 Seiten. 1974. DM 16,–

Vol. 368: R. J. Milgram, Unstable Homotopy from the Stable Point of View. V, 109 pages. 1974. DM 16,–

Vol. 369: Victoria Symposium on Nonstandard Analysis. Edited by A. Hurd and P. Loeb. XVIII, 339 pages. 1974. DM 26,–

Vol. 370: B. Mazur and W. Messing, Universal Extensions and One Dimensional Crystalline Cohomology. VII, 134 pages. 1974. DM 16,–

Vol. 371: V. Poenaru, Analyse Différentielle. V, 228 pages. 1974. DM 20,-

Vol. 372: Proceedings of the Second International Conference on the Theory of Groups 1973. Edited by M. F. Newman. VII, 740 pages. 1974. DM 48,-

Vol. 373: A. E. R. Woodcock and T. Poston, A Geometrical Study of the Elementary Catastrophes. V, 257 pages. 1974. DM 22,-

Vol. 374 S. Yamamuro, Differential Calculus in Topological Linear Spaces. IV, 179 pages. 1974. DM 18,-

Vol. 375: Topology Conference 1973. Edited by R. F. Dickman Jr. and P. Fletcher. X, 283 pages. 1974. DM 24,-

Vol. 376: D. B. Osteyee and I. J. Good, Information, Weight of Evidence, the Singularity between Probability Measures and Signal Detection. XI, 156 pages. 1974. DM 16.-

Vol. 377: A. M. Fink, Almost Periodic Differential Equations. VIII, 336 pages. 1974. DM 26,-

Vol. 378: TOPO 72 - General Topology and its Applications. Proceedings 1972. Edited by R. Alò, R. W. Heath and J. Nagata. XIV, 651 pages. 1974. DM 50,-

Vol. 379: A. Badrikian et S. Chevet, Mesures Cylindriques, Espaces de Wiener et Fonctions Aléatoires Gaussiennes. X, 383 pages. 1974. DM 32,-

Vol. 380: M. Petrich, Rings- and Semigroups. VIII, 182 pages. 1974. DM 18,-

Vol. 381: Séminaire de Probabilités VIII. Edité par P. A. Meyer. IX, 354 pages. 1974. DM 32,-

Vol. 382: J. H. van Lint, Combinatorial Theory Seminar Eindhoven University of Technology. VI, 131 pages. 1974. DM 18,-

Vol. 383: Séminaire Bourbaki - vol. 1972/73. Exposés 418-435. IV, 334 pages. 1974. DM 30,-

Vol. 384: Functional Analysis and Applications, Proceedings 1972. Edited by L. Nachbin. V, 270 pages. 1974. DM 22,-

Vol. 385: J. Douglas Jr. and T. Dupont, Collocation Methods for Parabolic Equations in a Single Space Variable (Based on C^1-Piecewise-Polynomial Spaces). V, 147 pages. 1974. DM 16,-

Vol. 386: J. Tits, Buildings of Spherical Type and Finite BN-Pairs. IX, 299 pages. 1974. DM 24,-

Vol. 387: C. P. Bruter, Eléments de la Théorie des Matroïdes. V, 138 pages. 1974. DM 18,-

Vol. 388: R. L. Lipsman, Group Representations. X, 166 pages. 1974. DM 20,-

Vol. 389: M.-A. Knus et M. Ojanguren, Théorie de la Descente et Algèbres d' Azumaya. IV, 163 pages. 1974. DM 20,-

Vol. 390: P. A. Meyer, P. Priouret et F. Spitzer, Ecole d'Eté de Probabilités de Saint-Flour III - 1973. Edité par A. Badrikian et P.-L. Hennequin. VIII, 189 pages. 1974. DM 20,-

Vol. 391: J. Gray, Formal Category Theory: Adjointness for 2-Categories. XII, 282 pages. 1974. DM 24,-

Vol. 392: Géométrie Différentielle, Colloque, Santiago de Compostela, Espagne 1972. Edité par E. Vidal. VI, 225 pages. 1974. DM 20,-

Vol. 393: G. Wassermann, Stability of Unfoldings. IX, 164 pages. 1974. DM 20,-

Vol. 394: W. M. Patterson 3rd, Iterative Methods for the Solution of a Linear Operator Equation in Hilbert Space - A Survey. III, 183 pages. 1974. DM 20,-

Vol. 395: Numerische Behandlung nichtlinearer Integrodifferential- und Differentialgleichungen. Tagung 1973. Herausgegeben von R. Ansorge und W. Törnig. VII, 313 Seiten. 1974. DM 28,-

Vol. 396: K. H. Hofmann, M. Mislove and A. Stralka, The Pontryagin Duality of Compact O-Dimensional Semilattices and its Applications. XVI, 122 pages. 1974. DM 18,-

Vol. 397: T. Yamada, The Schur Subgroup of the Brauer Group. V, 159 pages. 1974. DM 18,-

Vol. 398: Théories de l'Information, Actes des Rencontres de Marseille-Luminy, 1973. Edité par J. Kampé de Fériet et C. Picard. XII, 201 pages. 1974. DM 23,-

Vol. 399: Functional Analysis and its Applications, Preceedings 1973. Edited by H. G. Garnir, K. R. Unni and J. H. Williamson. XVII, 569 pages. 1974. DM 44,-

Vol. 400: A Crash Course on Kleinian Groups - San Francisco 1974. Edited by L. Bers and I. Kra. VII, 130 pages. 1974. DM 18,-

Vol. 401: F. Atiyah, Elliptic Operators and Compact Groups. V, 93 pages. 1974. DM 18,-

Vol. 402: M. Waldschmidt, Nombres Transcendants. VIII, 277 pages. 1974. DM 25,-

Vol. 403: Combinatorial Mathematics - Proceedings 1972. Edited by D. A. Holton. VIII, 148 pages. 1974. DM 18,-

Vol. 404: Théorie du Potentiel et Analyse Hormonique. Edité par J. Faraut. V, 245 pages. 1974. DM 25,-

Vol. 405: K. Devlin and H. Johnsbråten, The Souslin Problem. VIII, 132 pages. 1974. DM 18,-

Vol. 406: Graphs and Combinatorics - Proceedings 1973. Edited by R. A. Bari and F. Harary. VIII, 355 pages. 1974. DM 30,-

Vol. 407: P. Berthelot, Cohomologie Cristalline des Schémas de Caracteristique $p > o$. VIII, 598 pages. 1974. DM 44,-

Vol. 408: J. Wermer, Potential Theory. VIII, 146 pages. 1974. DM 18,-

Vol. 409: Fonctions de Plusieurs Variables Complexes, Séminaire François Norguet 1970-1973. XIII, 612 pages. 1974. DM 47,-

Vol. 410: Séminaire Pierre Lelong (Analyse) Année 1972-1973. VI, 181 pages. 1974. DM 18,-

Vol. 411: Hypergraph Seminar. Ohio State University, 1972. Edited by C. Berge and D. Ray-Chaudhuri. IX, 287 pages. 1974. DM 28,-

Vol. 412: Classification of Algebraic Varieties and Compact Complex Manifolds. Proceedings 1974. Edited by H. Popp. V, 333 pages. 1974. DM 30,-

Vol. 413: M. Bruneau, Variation Totale d'une Fonction. XIV, 332 pages. 1974. DM 30,-

Vol. 414: T. Kambayashi, M. Miyanishi and M. Takeuchi, Unipotent Algebraic Groups. VI, 165 pages. 1974. DM 20,-

Vol. 415: Ordinary and Partial Differential Equations, Proceedings of the Conference held at Dundee, 1974. XVII, 447 pages. 1974. DM 37,-

Vol. 416: M. E. Taylor, Pseudo Differential Operators. IV, 155 pages. 1974. DM 18,-

Vol. 417: H. H. Keller, Differential Calculus in Locally Convex Spaces. XVI, 131 pages. 1974. DM 18,-

Vol. 418: Localization in Group Theory and Homotopy Theory and Related Topics Battelle Seattle 1974 Seminar. Edited by P. J. Hilton. VI, 171 pages. 1974. DM 20,-

Vol. 419: Topics in Analysis - Proceedings 1970. Edited by O. E. Lehto, I. S. Louhivaara, and R. H. Nevanlinna. XIII, 391 pages. 1974. DM 35,-

Vol. 420: Category Seminar. Proceedings, Sydney Category Theory Seminar 1972/73. Edited by G. M. Kelly. VI, 375 pages. 1974. DM 32,-

Vol. 421: V. Poénaru, Groupes Discrets. VI, 216 pages. 1974. DM 23,-

Vol. 422: J.-M. Lemaire, Algèbres Connexes et Homologie des Espaces de Lacets. XIV, 133 pages. 1974. DM 23,-

Vol. 423: S. S. Abhyankar and A. M. Sathaye, Geometric Theory of Algebraic Space Curves. XIV, 302 pages. 1974. DM 28,-

Vol. 424: L. Weiss and J. Wolfowitz, Maximum Probability Estimators and Related Topics. V, 106 pages. 1974. DM 18,-

Vol. 425: P. R. Chernoff and J. E. Marsden, Properties of Infinite Dimensional Hamiltonian Systems. IV, 160 pages. 1974. DM 20,-

Vol. 426: M. L. Silverstein, Symmetric Markov Processes. IX, 287 pages. 1974. DM 28,-

Vol. 427: H. Omori, Infinite Dimensional Lie Transformation Groups. XII, 149 pages. 1974. DM 18,-

This series aims to report new developments in mathematical research and teaching – quickly, informally and at a high level. The type of material considered for publication includes:

1. Preliminary drafts of original papers and monographs
2. Lectures on a new field, or presenting a new angle on a classical field
3. Seminar work-outs
4. Reports of meetings, provided they are
   a) of exceptional interest and
   b) devoted to a single topic.

Texts which are out of print but still in demand may also be considered if they fall within these categories.

The timeliness of a manuscript is more important than its form, which may be unfinished or tentative. Thus, in some instances, proofs may be merely outlined and results presented which have been or will later be published elsewhere. If possible, a subject index should be included. Publication of Lecture Notes is intended as a service to the international mathematical community, in that a commercial publisher, Springer-Verlag, can offer a wider distribution to documents which would otherwise have a restricted readership. Once published and copyrighted, they can be documented in the scientific literature.

Manuscripts

Manuscripts should comprise not less than 100 pages.
They are reproduced by a photographic process and therefore must be typed with extreme care. Symbols not on the typewriter should be inserted by hand in indelible black ink. Corrections to the typescript should be made by pasting the amended text over the old one, or by obliterating errors with white correcting fluid. Authors receive 75 free copies and are free to use the material in other publications. The typescript is reduced slightly in size during reproduction; best results will not be obtained unless the text on any one page is kept within the overall limit of 18 x 26.5 cm (7 x 10½ inches). The publishers will be pleased to supply on request special stationery with the typing area outlined.

Manuscripts in English, German or French should be sent to Prof. A. Dold, Mathematisches Institut der Universität Heidelberg, 69 Heidelberg/Germany, Tiergartenstraße, Prof. B. Eckmann, Eidgenössische Technische Hochschule, CH-8006 Zürich/Switzerland, or directly to Springer-Verlag Heidelberg.

Springer-Verlag, D-1000 Berlin 33, Heidelberger Platz 3
Springer-Verlag, D-6900 Heidelberg 1, Neuenheimer Landstraße 28–30
Springer-Verlag, 175 Fifth Avenue, New York, NY 10010/USA

ISBN 3-540-07013-3
ISBN 0-387-07013-3